A Periodic Tale

My sciencey memoir

Dr Karl

Kruszelnicki

ABC
BOOKS

The ABC 'Wave' device is a trademark of the Australian Broadcasting Corporation and is used under licence by HarperCollins*Publishers* Australia.

HarperCollins*Publishers*
Australia • Brazil • Canada • France • Germany • Holland • India
Italy • Japan • Mexico • New Zealand • Poland • Spain • Sweden
Switzerland • United Kingdom • United States of America

HarperCollins acknowledges the Traditional Custodians
of the lands upon which we live and work, and pays respect
to Elders past and present.

First published on Gadigal Country in Australia in 2024
by HarperCollins*Publishers* Australia Pty Limited
ABN 36 009 913 517
harpercollins.com.au

Copyright © Karl Kruszelnicki 2024

The right of Karl Kruszelnicki to be identified as the author of this work has been asserted by him in accordance with the *Copyright Amendment (Moral Rights) Act 2000*.

This work is copyright. Apart from any use as permitted under the *Copyright Act 1968*, no part may be reproduced, copied, scanned, stored in a retrieval system, recorded, or transmitted, in any form or by any means, without the prior written permission of the publisher. Without limiting the author's and publisher's exclusive rights, any unauthorised use of this publication to train generative artificial intelligence (AI) technologies is expressly prohibited.

A catalogue record for this book is available from the National Library of Australia

ISBN 978 0 7333 4034 5 (hardback)
ISBN 978 1 4607 1168 2 (ebook)

Cover and internal design: Design by Committee
Front cover photography © Steve Baccon
Endpapers photography © Luisa Brimble
Illustrations by Lisa Reidy; images on pages 233 and 404 courtesy of iStock
All photos courtesy of Karl Kruszelnicki unless otherwise stated
Typeset in Bembo MT Std by Kirby Jones
Printed and bound in Australia by McPherson's Printing Group

This book is dedicated to my family, whom I love above all else:

My parents in the past, strangers in a strange land, who gave me their all;
   And now my own extended family, rich and wonderful with children, children's children, and multiple relatives of every kind spanning from the past, the present, and into the future.
   I treasure every generously shared experience.

Contents

An idiosyncratic introduction ... ix

Part 1: The young, shy, curious years ... 1
1948–50: War, my parents and me ... 5
1950–53: Coming to Australia ... 21
1954–64: School years ... 39
1964–65: Digging ditches ... 64
1965–67: Coffee, politics and uni – in that order! ... 70
1968–69: The steelworks ... 81

Part 2: The drug-crazed hippie years ... 103
1970–71: A hippie in New Guinea ... 107
1971.5–75: Budding filmmaker ... 134
1971.5–75: Roadie ... 145
1971.5–75: Knight of the Road ... 157
1971.5–75: Disturb the universe ... 189
1975–83: Glebe squats ... 204

1976–77: Hooroo, hippie! Hello, hospital! 226
1978–80: Me and eye 256

Part 3: The career-hopping, family-focused years 279
1980s: Love, loss and launches 283
1988–92: Why I became a doctor 320
1990s: Pregnancy and parenting 335
1987–90 and 1990–2010: Outback odysseys 345
1990–91: Working up a storm 354
1994–96: Julius Sumner Miller fellowship,
    horse racing and Erdős numbers 363
2003: The god of overcoming obstacles and
    new beginnings 378
2006: Getting hitched 399

Epilogue 411
Acknowledgments 419

An idiosyncratic [In]troduction

Usually, I tell stories about new knowledge. But in this book, the stories are about me, and about the sidesteps, the U-turns, the outright failures and the good old dumb luck that made me who I am today!

So, here's some things about me, to get you in the mood. Don't worry, there's no exam!

I have seven names because of weird Polish heritage stuff. I have an asteroid named after me (18412 Kruszelnicki), but my true claim to fame was that I helped birth the word 'selfie'. In 2003, Nathan Hope posted a blurry photo of his swollen and sutured upper lip, and asked why the wound was so itchy. He typed, 'Sorry about the focus, it was a selfie.' According to the *Oxford English Dictionary*, this was the first time the word was ever written down. And where did it appear? Yep, on the Dr Karl Self Service Science forum – my web page!

Dr Karl Kruszelnicki

> **Kr**
>
> The first two letters of Kruszelnicki are Kr – and in the periodic table, they stand for Krypton (which is a very unreactive gas). Krypton (discovered in 1898) precedes Superman (1938) by four decades. By a 'coincidence', Superman comes from the fictional planet Krypton. Inexplicably, 'kryptonite' (from his home planet) can drain all of Superman's powers – and worse.

People think I've got a good memory, but I don't really. I have a trick. I make stories out of everything to lock the knowledge into my head. It seems the human brain is uniquely wired to remember stories. Telling stories is part of social bonding. Just look at how much of your time is spent sharing or hearing stories. Maybe this is part of making us form strong groups to compensate for our pretty hopeless physical abilities.

I use this storytelling trick to turn all my scientific readings from bare facts into true tales, with a beginning, a middle and an end, that I can tell anyone who happens to have a sense of curiosity. This helps me remember each one of my stories as a single and complete parcel of information, with the numbers and facts organically embedded in the storyline.

As part of my creative process, I try to turn scientific jargon into street talk. This means that you (the reader/listener) don't need any specific background science knowledge. And because it's a story, you can easily remember and retell it (if it tickles your fancy).

This book is based almost entirely on my memories, so there's only me to blame if any of it's wrong. After all, the easiest person to fool is yourself, and false memories are very common! Some parts of this book might be absolutely true, and some parts might not be true at all.

Take this memory as an example. I have a very clear image of driving around Melbourne in the early 1970s, when I was a drug-crazed hippie, in my gutless rustbucket of a brown-and-white VW Kombi van. As I cruised slowly around a corner, my friend in the back lost his balance and fell against the side door, which burst open, and he fell out onto the road! Luckily, he wasn't hurt. That's my memory. But his girlfriend at the time has a completely different memory of this same event: I was not the driver, and she was the one who fell onto the road! At least one of us, and maybe neither of us, has no idea of what actually happened.

People also tell me I'm super smart, but I'm not. I've worked with really smart people in lots of workplaces, and I know I'm not one of them. They have ideas and insights that I don't. However, recognising a brilliant idea (and sharing it) is one of my special skills.

Sometimes people say, 'Dr Karl, you know everything.' I don't! I'm a generalist with broad interests. The reason I know a lot is because I've had twenty-eight years of education in total, and I'm still reading and learning. I've spent sixteen whole years at university. I was lucky to be educated at the 'right' time in the distant past when the Australian government saw education as a worthwhile investment in the future. So I thank the Australian taxpayers (over and over) for my education.

I am genuinely interested in absolutely everything. Why is something this way but not that way? Why is the sky blue instead of red? If grass runs on sunlight, why isn't grass black to absorb as much sunlight as possible? (Overheating is a possible answer.)

I'm always curious, and I never mind admitting 'I don't know'. I happily ask, 'Can you explain that to me?' Remarkably often, people will share their knowledge, and I can get all their hard-won insights in a nice, neat summary.

I talk to everybody, everywhere. I never know when to stop asking questions! That's another reason why I know 'stuff', because I'm always seeking answers.

I don't understand why people worry about being wrong. If you don't make a mistake, you don't make anything. (Just try not to make the same mistake over and over.)

I started reading books from the local library in Wollongong (where I grew up) around the age of seven. I began with fairy tales, and then I accidentally discovered science fiction and was hooked. I devour science fiction. Averaged out, I read one sci-fi story every day of my life from age twelve to thirty-two.

I taught myself to speed read at a thousand words per minute to consume as many stories as possible. Sure, I missed some of the subtleties, but I still got the mind-blowing 'sciencey' concepts. The big ideas were what I was after!

Science-fiction books are better than sci-fi movies because both the storyline and the pictures in your head are superior. Sometimes sci-fi movies are just too lazy, with lousy scripts set in a dystopian future where it rains a lot.

I cut back on my sci-fi reading when I got to age thirty-two because I started studying medicine. There just wasn't enough time to load up my brain with everything I needed to become a doctor. But after that window of intense study, I dived back into sci-fi books again.

I love short shorts and I'm short-sighted – but they're not related to each other at all. I remember very clearly the discovery of my poor eyesight. I was sitting at the kitchen table and said to my mother across the room, 'I can't see you clearly.' My short-sightedness was almost certainly due to reading too much (sci-fi) under poor light with no breaks.

Why reading too much as a kid causes short-sightedness

First, some basic anatomy.

You have two eyes. Each eye is close to spherical, about 24 millimetres in diameter (much smaller than a golf ball, which is about 42 millimetres). The eye starts off about 16 millimetres in diameter at birth and then grows rapidly, eventually reaching full size around twelve years of age. The retina is an inner layer covering about three-quarters of the globe of the eye at the back and is about 0.3 millimetres thick. (Yep, like the Earth, the eye is a globe!)

Reading a lot in poor light when you're young can set you up for having to wear glasses. There is no problem for your eyesight when you stare vaguely at the outside world – the image will land on the retina without any specific focus point. But when you focus precisely on something small (like the words on a page), you will swivel your eyeball so that the words land on a specific sunken part of your retina called the fovea. The muscles around the lens of your eye adjust to bring the image into sharp focus on the fovea.

The fovea is very important in sharp vision. Compared to the entire retina, the fovea is tiny – only about 1.5 millimetres across. But around half of all the information that goes to your visual cortex to be turned into glorious 3D colour vision comes from that tiny pit.

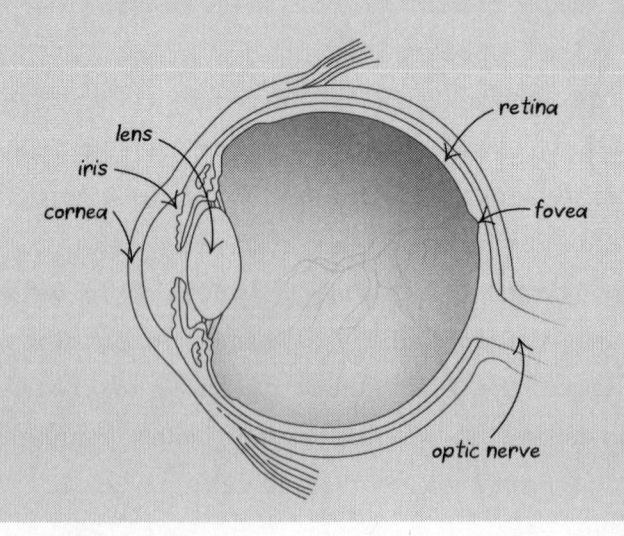

> The problem is the sunken pit: if the image is in focus in the fovea, it will be out of focus on all of the rest of the retina. Bummer! This is because the fovea is a bit further away from the cornea than the rest of the retina.
>
> Suppose you're doing a lot of reading in the years before you hit twelve (when your eye stops growing). If you read a lot in low light, the eye grows bigger in a fruitless attempt to get a sharp image all over the retina – not just the fovea. If you do this for many continuous hours each day and for many days in a row – like, if you're reading hundreds of sci-fi books at a thousand words a minute – you will grow a (slightly) bigger eyeball. And those bigger eyeballs are more likely to need glasses for long-distance vision.

At primary school, I 'invented' wearing socks with sandals. This gave me the protection of wearing shoes without the sweaty feet. But even in primary school, my fellow students were sceptical about my personal style!

I totally believe in the Rainbow Colour Theory of Dressing. In other words, I wear all the colours of the rainbow all at the same time. Remember that weird saying 'blue and green should never be seen without another colour in between'? Take no notice and let your imagination run free. And if I'm ever dressed in all black, you can be sure aliens have taken control of me!

For many years, all my clothes came from op shops. My mum used to volunteer at St Vincent de Paul and brought home things that caught her eye. I wore them happily. Her most magnificent find, which I still love to wear, is a leather vest handpainted with flowers. It's a beauty! Now my wife makes all my colourful, playful shirts.

I waited until the relatively grand age of twenty-eight years before I passed two major Australian milestones: learning to swim and learning how to eat a mango.

I should start off every conversation with an apology, because I regularly (but not deliberately) upset people by not reading the room (and being too concrete and literal).

I love stainless steel sooooooo much. It's a miracle. Hey, I'm joking! It's not really a miracle – it's chemistry. Stainless steel has lots of chromium (about 18 per cent) and nickel (about 8 per cent), which makes a shiny stable 'rust' that is both non-porous and adherent. I suspect that some varieties of stainless steel (there are hundreds) will outlast the Pyramids.

I love glass containers too – they are perfect for storing food. Glass should also outlast the Pyramids.

I am a very cheap drunk – three or four glasses leave me legless. Along the way I lose the gift of speech, and get reduced to being only able to repeat the last three or four words of the conversation.

I'm not really a sports fan – except for the Sydney Swans Aussie Rules team. Alice, my daughter, is a true fan who can recognise the Swans players by their arm muscles alone! I am impressed by her highly trained eye, and I really respect the skills of the athletes too.

In first-year medicine, I shaved off half my beard and kept it that way for several weeks.

Steamed soya beans and vegetables with tahini are my favourite foods (but I still crave, and eat, meat at times). They are truly a very nutritious and extremely delicious combination (according to me, not my family). I get confused when my children reject steamed cucumber for their dinner. That's totally my bad, because zucchinis and cucumbers look very similar to me!

Sometimes it takes me a long time to make a decision. I have to think about every single scenario, and my brain gets so busy running through all the possibilities that sometimes I don't have enough brainpower left over to speak as well.

I love working on a car. If everything 'goes to shit', there is always the option of jacking up the windscreen wipers and rolling a fresh car underneath.

My very favourite household obsession is stacking both dishwashers as carefully and efficiently (though some would say rigidly) as possible. Two dishwashers? Under high-use conditions, you can load one dishwasher and run it, empty it straight to the table, and then load the dirty dishes back to the other dishwasher – and repeat. Ideally you can bypass having to put the plates back in the cupboard at all!

Hanging washing on the old-fashioned rotating Hills Hoist clothesline, *and* in my very specific way, is another delight. Big heavy things (towels, jeans etc.) go on the outermost line so they get the most sunlight. As you move towards the centre, the clothes on each line are pegged according to their size: big (jeans), little (undies), big (T-shirts) and so on until you reach the centre. This maximises drying. Undies always get turned inside out so the ultraviolet from the sun can kill germs. I dream of installing a force plate on the vertical shaft of the hoist that will measure the minute-by-minute loss of weight as the water evaporates. Maybe a project for when I get some spare time?

I think that some actions should be locked-in reflex behaviours with no variation allowed. These include always stopping at a red light or stop sign, always using the blinkers to change lanes while driving, and always wearing a seat belt when in a car. These conditioned behaviours could save your life! Each year, about two million motor vehicle collisions happen because the driver did not use their indicators.

I try to do things immediately as I think of them so I don't forget later, but this makes me very distractable.

I'm really lateral-minded, and I'm easily drawn to shiny new things.

I never had a long-term plan for how my life should turn out. I never looked from the top of the metaphorical 'tall building' to see all the possible paths my life could follow and choose the best one. Lucky for me, things came out fine in the wash! I've lived as if I were an ice-block stick floating in the gutter of life on a rainy day. (Thanks, Norman Gunston, and sorry if I'm misquoting you – blame my bad memory!)

I just love the way that science is always finding out new stuff. There will always be amazing things to be excited about. I never, never, never want to stop learning about the latest discoveries in science and passing them on.

I'm hoping this book brings my sense of joy, awe and wonder at the universe into your brain. Two things for some peace of mind. First, the universe is random (thank you, quantum mechanics!). Second, there is so much stuff we don't understand. Put together, they mean we don't have to have everything already figured out – we are (partially) off the hook and (partially) free of responsibility. We are all part of the tumultuous roller-coaster ride of life! You get on and off at the same place – but the ride in between is what matters.

So how do all these little bits of me add up to make one person, Dr Karl, journalist, family man, sciencey talkback guy, author, Mr Bright Shirts, and micro-star of stage and screen? Well, luck, mostly! Sure, I made decisions that led me to test four-wheel drives in the Australian outback for a few decades, work as a TV weather presenter and give up working as a kids' hospital doctor to instead bring science (and the truth about vaccines) to everyone. If I'd made different decisions, maybe I could have been an Olympic gold medallist for the high jump, a Nobel Prize winner or an early bitcoin investor and multibillionaire.

I'd be lying if I said I'd never thought about all those 'maybe' endings. But if I changed anything I would have a whole different box of frogs (or kettle of fish), and what I have now is practically perfect for me. So I'm at the happy point where I wouldn't change anything at all.

Sometimes, you're lucky …

Part 1

The young, shy, curious years

Both my parents were subjected to terrible and unimaginable horrors for a good chunk of their lives, until they ended up in a refugee camp in Australia. Before that, they had been in another refugee camp in Sweden. And before that, their lives teetered between being barely alive, and the ever-present threat of imminent death. They were shuttled like sacks of potatoes from ghetto to concentration camp, and between concentration camps – with very little control over their lives.

Despite all these odds against them, they fell in love in Sweden. A little later on, I was born. And very shortly after my birth, they felt threatened by a potential war between Russia and Finland. It was only a few hundred kilometres from Finland to Sweden – much too close for comfort.

So there I was in Sweden in 1950 – an only child, two years old, already a 'reffo' kid but with no idea yet about the true story of my parents' backgrounds, ready to cross the wide blue ocean to a brand-new life. As a little family of three, we were supposed to go to North America, but the Random Nature of the Universe intervened, and we ended up in faraway Australia.

The many and terrible traumas that my parents had been through affected them, and how they behaved, in very different ways. And their habits and histories affected me as I grew up. As a family, we had little in common with the Anglo Australians around us. But there were some things we shared. Just like my schoolmates, I was told to eat more, because of the 'starving children in Africa'.

After settling in Australia, I found a way through a childhood that could have been very lonely. I ended up morphing from a shy immigrant kid into a university graduate with a proper job, at the Port Kembla Steelworks in Wollongong.

So how did I get from Sweden to Wollongong?

1948–50

War, my parents and me

My first memory is weird, because for most people, seeing light at the end of the tunnel is supposed to be their last memory.

But I can remember being a tiny baby in my mother's arms on a railway platform and seeing light from a train in a tunnel heading towards me. The sky was jet black, but the streets were white and there were people everywhere. I'm guessing that this memory was from Sweden during the depths of winter, with snow on the streets.

Intellectually, I realise that it might not be a true memory at all, because in Western societies, people hardly remember anything from before the age of six. But in cultures where storytelling and communal gatherings are more common, people can have memories from well before the age of six.

Childhood amnesia

There are various guesses as to why you usually can't remember anything from when you were a baby.

One relates to the concept of *self*, which supposedly helps lock things into your memory. For example, when a baby can recognise themself in the mirror, they can also better remember how to play with their toys. This suggests that memory and ability to recognise self are somehow linked.

Another guess relates to *language*. Consider seven-year-old kids who attended emergency departments when they were two. They have better memories of what happened in the hospital if their language skills were pretty strong at age two. If they didn't have a lot of words at two, they remember less of what happened to them at the hospital.

The third guess relates to *size*. As an infant, you can only see the underneath of furniture. As you get older, the world is literally seen from a different angle compared to when you were smaller. Your infantile memories simply don't match your new world perspective, so perhaps they get ditched.

> The latest guess relates to *neurogenesis*, which is laying down new nerve cells. Babies lay down 700 new nerve connections each second. There is a balance between adding new information, and being able to keep old information. It is possible that part of the price of the rapid brain development (that happens in the first six or so years of life) is that you lose those very early memories.
>
> This leads to a very odd situation. Youngsters are such incredibly efficient sponges for information, and their first few years of life are very powerful in shaping their personalities. Yet surprisingly, the small details are almost all forgotten.
>
> So, parents of youngsters, don't take it personally when your grown-up kids can't remember the sleepless nights and blood, sweat and tears that you poured into raising them. Nothing personal – it's probably just a side-effect of how the brain grows most efficiently!

I was born in Sweden in 1948, just three years after the end of the Second World War. Our family lived with an 'adoptive' family in Helsingborg, the ninth largest city in Sweden, on the south coast and at the closest point to Denmark. As an aside, Helsingborg's sister city in Denmark is Helsingør (known in

English as Elsinore), where Shakespeare's Hamlet walked the castle battlements. Helsingborg is just a 4-kilometre ferry ride across the water from Helsingør.

The father of our Swedish adoptive family was called Karl, so my parents named me Karl Sven for him. But they never called me Karl when I was young – I was always called Wojciech (pronounced 'Voytek'). It wasn't till late high school that I switched from my very Polish name to my very Swedish name. To be fair, I had six names (Karl Sven Wojciech Sas Konkovitch Kruszelnicki) to choose from! (And I got a seventh name, Matthew, when I was confirmed in the Catholic Church.)

Supposedly, I was a very good baby who didn't ever cry. But I also remember my mother telling me that she used to pop me in the cot in the morning (in Sweden) before going to work (in Denmark), come home at lunchtime to feed me, and then head back to work. I'm not sure how she would have known if I was or wasn't crying – if that story is true! You wouldn't get away with that kind of parenting these days, but I guess people do what they have to do to survive.

Officially, Sweden was neutral during the Second World War. After the war it accepted a limited number of refugees from war-torn Europe, and my parents were among them. They met and fell in love in a Red Cross Displaced Persons Camp in Sweden, where my father was teaching English to fellow refugees. Apparently, my father used to push my mother around in a wheelbarrow, because she was still too physically weak to even walk in the aftermath of the war.

My father pre-1950

My father, Ludwik, was a Catholic Pole with a fabulously colourful life.

He graduated from law at a university in Lviv but spent only a short time working as a lawyer. He became disillusioned when the clients he represented successfully later told him that they were in fact guilty! He then did a lot of travelling through Egypt, Europe and the United States.

In Chicago he worked as a journalist for the *Chicago Tribune*, the modestly self-proclaimed World's Greatest Newspaper. (In 1924 it set up a radio station with the call sign WGN!) The *Tribune* was in a ratings war with the eight other newspapers in Chicago and chasing the big headline stories, which pushed my father into becoming very good at getting news copy out quickly.

On one occasion the boss came to him with a 'terrific' photo of a young woman who had died in a road accident in downtown Chicago – but absolutely no other information. The photo was going on the front page, and the presses would roll in fifteen minutes! (Today, that photo would be called click bait.) My father had to make up a story to go with the photo, pronto. He churned out a totally fictitious story about a woman who had mistakenly believed that her family had died in a tornado in Kansas. She moved to Chicago to start a new life, but that very morning she spotted her husband and two children just across the street! As she raced to reach them (oblivious to everything else) she was run over by a truck and killed! A classic tear-jerker.

He moved on to Hollywood, where his talents were clearly better suited. He got wealthy writing scripts, including one for a Three Musketeers movie.

Dad fought in the Spanish Civil War. He was Catholic and, somehow, thought it was his duty to fight the Communists. He loved Spain so much that he 'burned' the language deeply into his brain. Half a century later in New Caledonia, he spoke with a Spanish family who were also on holidays. He was so fluent that they were convinced he was a native Spanish speaker!

On another occasion he worked as an English translator for the Polish embassy in London, where he spoke the King's English with a faultless upper-class British accent.

In total he spoke twelve languages, including Classical Greek and Latin. Holidaying in Bali in the 1970s, my mother got sick, and we couldn't speak the local language to ask for help. My father found a Balinese Catholic priest and communicated in Latin (the language of the Catholic Church at that time) to work out how to get my mother medical care. Not bad for a 'dead' language!

By 1939 he had returned to his hometown of Lviv, which at that time was in Poland. When war broke out, the Soviet Union invaded Lviv. My father wrote an article against Communism, which made him an Enemy of the State. He was dobbed in by a fellow worker at the electrical plant and arrested by the NKVD, the Soviet Internal Affairs Security Agency. Waiting on a railway platform with a bunch of fellow prisoners who had all been sentenced to exile in Siberia (in plain English, to die in a hard-labour camp), the train was delayed. The Soviet guards took a break and just left the prisoners unattended. My father saw a chance and ran like crazy.

He headed north some 900 kilometres until he got to Lithuania, where he planned to take a boat to Sweden. His goal was to get to the United Kingdom and fight with the Polish forces there against the Nazis. Unfortunately, at the border he was apprehended and handed back to the local Soviets as a spy – two strikes!

He was held in Grodno prison for over a year and tortured as part of the usual practices of the NKVD. The treatment was harsh, and prisoners were deliberately sleep-deprived as part of the punishment. He was only ever referred to by a number, not by his name. Shockingly, the depersonalisation actually worked: he briefly forgot his parents' names but managed to retrieve them from somewhere in his memory. He also lost his English language skills. (He did learn English again in Australia after the war, but now he spoke it with a Polish accent, not the upper-class British accent of pre-war.) He was finally sentenced to exile in Siberia for fifteen years (again, to die in a hard-labour camp) on 8 June 1941.

Luckily for him, the actual paperwork took a little time to organise. Even more luckily, Nazi Germany declared war on the Soviet Union on 22 June and bombed his Soviet prison camp. The walls were breached and he escaped with a friend into the unknown. That night there were exactly fourteen stars to be seen in the heavens, so my father adopted fourteen as his lucky number for the rest of his life. (It was the number of our house in Wollongong, by total chance!)

He made his way overland back to Warsaw where he became active in the Polish underground, which gave him fake identity papers so he could travel around the city. I remember he told me that he smuggled people out of Nazi Poland, almost certainly Jews, which was thought unusual behaviour for a staunch Catholic. That made him a lot of enemies.

He moved in with an old girlfriend who was a medical doctor and a little 'volatile'. On one occasion before the war, she had shot him in the face at close range, but amazingly the bullet passed through his left cheek and out his open mouth. Maybe that should have been warning enough – but war makes for strange bedfellows.

One night at the theatre, this girlfriend left her handbag with him while she went to the bathroom. He opened it to look for a handkerchief and found a document declaring her to be an honorary Nazi German citizen. They had an argument, because in his eyes, this made her a de facto traitor. He told her she was on the losing side. The next morning, she left much earlier than usual for work. Shortly afterwards, the Gestapo turned up at the door and took Dad away.

(Weirdly, this dangerous ex-girlfriend managed to track down his address in Wollongong and wrote to him until the 1960s, asking him to come back to her in Poland! But he'd finally learned his lesson, and was happily and deeply in love with my mother, and me.)

Initially he was held in Krakow concentration camp as a political prisoner, then Auschwitz, and later he was transferred to Sachsenhausen in Germany's north-east. As he came off the transport train, one of the kapos (a fellow prisoner assigned to work for the Nazis) was whipping the prisoners without too much enthusiasm. As he got close to my father, he muttered under his breath, 'You are eighteen and you have a trade.' My father took his cue from the kapo and told the guard, 'I am twenty-five and I'm a carpenter.' He was actually forty years old, with no manual skills whatsoever! But he was strong and robust, so he was sent to a much shorter queue of prisoners. That decision once more saved his life, because the people in the longer queue were all killed very shortly after.

At one point my father's job was taking the bodies of prisoners out of the gas ovens to another area in the camp for cremation. After doing this for a while, he quite naturally got very depressed, and saw no hope of ever escaping the same fate. One day he was carrying yet another dead body when the corpse's head started rocking from side to side as he trudged. In his mind he started hearing a bird calling 'cuckoo, cuckoo' in time with the movement. It made him laugh involuntarily, and somehow this gave him the strength to keep hoping he might make it out alive. (I, too, have always shared his irrational optimism. I think that, on average, being optimistic will give you better results than being pessimistic.)

I have postcards he sent to his sister, back in Poland, from Sachsenhausen. They have a stamp featuring Hitler on the back! I get chills whenever I look at those cards. The messages he wrote to his sister have been translated for me and, incredibly, he never once mentioned his own trials. Every message was hopeful and concerned only about his family's well-being.

The almost inevitable likelihood was that one day, his number would be up. Even though the Allies had invaded

Europe and seemed very likely to overcome the Nazis, he had been a witness to their war crimes, and the Nazi practice was to ensure there were no witnesses. He survived because of three pieces of luck.

First, a Soviet prisoner of war (a military prisoner, not a political prisoner like my father) died of appendicitis in the Soviet section of the camp. (Of course, this was not lucky for the deceased soldier.) Second, as a good chess player, my father often played with the German medical camp doctor, and they built an odd kind of rapport. Third, my father had managed to hang on to something incredibly precious – a single tin of sardines – that he was able to use as a bribe.

With the help of the doctor, my father swapped identities with the dead Russian soldier with the burst appendix. The official camp records showed a body with the forearm tattoo number of 95808 (my father's number) was declared to have died of appendicitis.

My father told me that after he had been transferred to the Soviet section of the camp, he saw the Soviet prisoners killing and eating each other – cannibalism. There was very little fat or muscle left on their emaciated bodies, but the meat of the brain was preserved and relatively unaffected by starvation. He actually saw them cracking open the weaker prisoners' skulls to eat their brains.

Units from the Soviet and Polish armies liberated the camp shortly after he took on the Soviet prisoner's identity. Luckily his tattoo identified him as a political prisoner, so he didn't have to argue too hard about not being a real Soviet soldier. Somehow he ended up in a lazaretto (a military hospital) in

Lubeck, Germany from 2 May to 21 July 1945. Even more luckily, a random Red Cross worker asked my father if he wanted to go to Sweden into a displaced persons' camp. There was no hesitation. He immediately said, 'Yes, please!'

In late 2011, I visited the site of Sachsenhausen concentration camp with my whole family, and I was so glad of their emotional support and all of their hugs. I deliberately paced back and forth around the courtyard and assembly areas, trying to make sure my footsteps were retracing where my father's would have been, half a century earlier. The December winter wind was biting, and I was freezing. It was so easy to imagine how little warmth the thin cloth of the prisoners' uniforms would have offered, but actually, the cold would have been a minor part of their suffering. Tears just kept pouring uncontrollably out of my eyes.

The stories that my father told about his wartime experiences were so vivid and wild. At the time, I didn't know if the stories were pure theatre, or true events with a journalistic spin, or simply the stuff of real nightmares. I thought he was spinning a yarn, but historians have confirmed his life story for me. The experience of being on the SBS television series *Who Do You Think You Are?* really opened my eyes to my parents' lives. I saw many official documents with my parents' names, and sometimes signatures, from their lives in Europe that validated my father's stories. It deeply affected me, because the depth and reality of their suffering was undeniable.

My father died in 1981. He would have shared even more of his experiences with me, but I was stupidly foolish to not realise what I would lose by not talking with him. Luckily I have kept all of his professional writings for various European journals that he wrote when he lived in Australia. He usually wrote about Australiana – but sometimes he wrote of his past, and other times about his family living in the strange exotic land Down Under. My son, Karl (yep, same name), had his articles translated from Polish into English for me as a present. It was a beautiful and meaningful gift that again opened my eyes. It's an odd experience reading stories about me from the past that my father wrote.

Ultimately, the message from my father that stuck with me all my life was that you have to be optimistic to survive, and life is a precious gift that you never want to give up on.

My mother pre-1950

The contrast between my parents was huge: my father wanted to tell me too much about his ordeals, but my mother wanted to tell me nothing at all.

I was confused about my mother's background for a long time. Sometimes she said that she was born in Sweden and was Lutheran. At other times she said that she was a Polish refugee from Gdansk and met my father in a Red Cross camp in Sweden after the war. She was long dead before I finally learned her true story.

My mother, Rina, was a Polish Jew. Tragedy had filled her life from early childhood. When she was only one, her father was struck by lightning while out hunting in the forest. He had sheltered in a small hut, but that did not protect him. The lightning bolt struck and killed him.

Rina's mother remarried some ten years after he died, and she later had two more children. During the war, my mother was in the Łódź ghetto, the second largest in German-occupied Europe; some 200,000 people passed through it. Because she was an older woman, Rina's mother was transferred on a train with the very old and the very young early in the piece. She almost certainly died soon after.

Part of the reason that the Łódź ghetto survived for so long in the war was that the inmates organised themselves into a working ghetto, and were therefore 'useful' to the Nazis. They did a lot of work with clothes. I cannot imagine what those who had not been sent to the death camps felt when the clothes of those who had been sent away then turned up for them to process and repair. They must have known that their fellow ghetto citizens were now dead.

I've been to Łódź and seen the memorial tree number 626 planted in my mother's memory in the Park of the Rescued (commonly called Survivors' Park), which is dedicated to survivors of the ghetto. Seeing her tree helped me to unravel some of my own roots – and again, I couldn't avoid crying.

My mother ended up in Auschwitz concentration camp and survived by the skin of her teeth. She told me that sometimes they would see Allied planes flying high overhead. They actually prayed for the planes to drop bombs on them, to give them a quick and merciful death.

One morning they woke up to find that all the Nazi guards had vanished. A few hours later they were liberated by the Soviet army. My mother had tried to protect her younger sister, Stella, for the whole length of the war inside Auschwitz, but they were separated during the liberation. My mother searched unsuccessfully for Stella for years and finally found out that she died in the 're-feeding'. The bodies of many of the starving prisoners were simply unable to manage the rich food supplied by the well-meaning liberation forces. (In comparison, my father told me he had only the water that rice had been boiled in for the first day of liberation, only rice on the second day, and added meat after that. I don't know how he knew that he needed to slowly retrain his body to accept food again.)

My mother never saw any of her birth family again.

Rina told me barely anything truthful about her past. She lied to cover up her Jewish heritage and her wartime experiences, and despite her heavy middle-European accent, I blindly accepted her stories, even when they contradicted one another. I really didn't think about the discrepancies much at all. She lied to me about her religion, her place of birth (she was actually born in Łódź in Poland) and her age. On entering Auschwitz, she lowered her year of birth by eight years, from 1914 to 1922, so she could be interned in the same group as her younger sister. She even lied about her name (she was christened Ryvka, not Rina). All these discrepancies made it incredibly difficult to track her past, and maybe that was the idea. She maintained these fictions on all official documentation; on my Swedish birth certificate, it incorrectly records my mother as being born in Gdansk in 1922.

When I was an adult, my mother slowly revealed little pieces of her real life to me. Sadly, it was with the onset of dementia towards the end of her life that she spoke most openly about things she had seen in the war. She was haunted by memories of dead bodies. Her imagination saw these war ghosts piled up in our courtyard at home. She feared even more intensely who was 'with her' and who was 'against her'.

For much of my life, I kept some emotional distance from my mother. Her distrust of society at large was hard for me to accept, but now I can understand why she was so constantly suspicious of the outside world. I remember her saying to me over and over, 'You don't know people.' She was probably right, and I do know how lucky I was that I had never known people the way she had in the war.

One thing was always clear, however: she absolutely idolised my father and put him on a pedestal her whole life.

Hit the road

Like other refugees, my parents had tremendous physical and emotional scars from their wartime experiences. That terrible suffering, along with the fact that the Soviet Union was making very aggressive noises at nearby Finland in the immediate post-war era, meant they wanted to get as far away as possible from that part of the world. My father had already been captured by the Soviets and tortured in a brutal Soviet prison for over a year, so he really didn't want to 'enjoy their hospitality' again. My mother was also quite determined to be as far away as she could be from any potential war, and all her heart-wrenching memories.

So in early 1950, with me just two years old, they decided to leave Sweden and Europe for good. My parents packed everything we owned into cheap cardboard suitcases and did all the paperwork required to apply for immigration to 'Anywhere Else' in the world.

In April 1950 my father made an official application to the International Refugee Organization and filled out section 22, 'Reasons for Not Returning Home', to explain why he should *not* be sent from Sweden back to Ukraine:

> I was persecuted and sentenced by Soviets NKWD to 15 years jail. On the 22nd of Juni 1941, I succeeded in escaping. It was the first day of the German-Russian War and our jail at Grodno has been bombarded by German AF. My home Lwow is now Soviet Russia (and behind Curzon Line) and the present Poland and Polish Police is now in close collaboration with Soviet NKWD (now MWD).

My parents both recorded their nationalities as 'Stateless' — this was (and still is) the sad reality of life as a refugee.

In another section they had to declare everywhere they had lived in the last twelve years. Their answers read like a long list of anywhere you never want to go! They had been moved between concentration camps, death camps, ghettos, Soviet jails and hospitals. Not what you would call the height of luxury …

When this is your resume, you may as well hit the road! They headed to the dock with a plan to emigrate to North America.

The road for them began with thousands of kilometres of big blue ocean.

1950–53

Coming to Australia

We had planned to go to North America, but an unexpected vaccine side effect turned out to be all for the best!

We went to the dock intending to get onto our ship, but just before we boarded, I spiked a fever from my compulsory smallpox vaccination. Vaccines do have common minor side effects, which are nearly always less serious than the disease itself. (The very nasty disease of smallpox no longer infects humans, thanks to vaccinations.)

My parents freaked out. I was their only child! They decided to play it safe and stay on dry land with their sick baby in case I got worse, so the ship sailed away to North America without us. And just as we waved goodbye to that ship, my fever sailed away too. According to my parents, there we were, still sitting on the dock with our cardboard suitcases, when the next ship berthed in that dock. It was going to Australia – and we simply boarded. The ship was called SS *Goya*.

So that's how, thanks to vaccination, I randomly ended up in Australia.

I can't remember anything about the trip. It must have taken about a month to travel the 15,000-plus kilometres all the way from Sweden to Australia. It would have been a time of mixed feelings for my parents, including great apprehension about their new life and new home. But for them, one huge upside was that they would be a long, long way away from Russia – even further than being in North America!

(Being a great distance from a war is a good thing. To put the sense of safety that they got from that distance into perspective, think back to the potential Russian invasion of Finland. If they stayed in Sweden, they could have been just a few hundred kilometres from a military front – roughly the distance between Sydney and Newcastle!)

We arrived in Australia as part of more than two million displaced people fleeing war-torn Europe in the two decades after the end of the Second World War. The official Australian government 'White Australia Policy' was shifting to a 'populate or perish' mindset. But while the government was opening the doors for immigration post-war, not all the locals liked the change of policy.

After docking in Australia, we ended up in a refugee camp on the border of New South Wales and Victoria, checking into Bonegilla Migrant Reception Centre on 8 November 1950. I honestly don't remember anything much about that time of my childhood, but I went back a few years ago to see the little hut the three of us lived in. It was pretty small, about the size of a tradie's van. My parents told me we were provided a ration of eggs, which they always gave me, to make me stronger. That story about their willingness to sacrifice their own share of food for me also brings me to tears. I guess it's what most parents would do, but it touches me so deeply that even though they had so little, they were willing to give so much to me.

During our time at Bonegilla, our little family of three was temporarily broken up when my father left after three weeks to work for the Metropolitan Water Sewerage and Drainage Board. The Water Board serviced all of metropolitan Sydney as well as the Blue Mountains in the west and Wollongong to the south. My father was stationed at Regents Park, some 32 kilometres west of central Sydney.

My mother and I left Bonegilla a few months later in February 1951. We ended up on the other end of Sydney in the Northern Beaches, roughly 30 kilometres north of central Sydney. My mother was the housekeeper for a doctor at Collaroy and I remember him and his family, who sponsored us, as being lovely. They had a gorgeous house on Pittwater Road just across from the ocean. We stayed in touch with our sponsor family long after we left, and they visited us every Easter for many years to come.

I was in childcare near Collaroy. I must have had migration in my blood at that time because one day when the door to the centre was accidentally left open, I walked my three-year-old self out the door. My mother and the doctor found me toddling down the centre strip of the busy divided main road all by myself. I can remember my mother picking me up and crying and crying without stopping. I don't think I would have known what all the fuss was about at the time, but I remember her absolute distress that I could have been hurt, or worse.

At the beginning, the best job my father could get was labouring, despite his language skills and professional training. He was fine with doing any work. He was good at the labouring work and did it easily, competently and honestly. He stayed with the Water Board for the rest of his working life.

It's still a common story for migrants and refugees in Australia to take on unskilled work, despite professional qualifications in their own country. Even today, I see this waste of brainpower and skills. All too often, a taxi or Uber driver who grew up overseas tells me they have a PhD in a field that is obviously economically important to the future of Australia. But then they tell me they can't get work in Australia in their field. It seems we still haven't really learned how to value the talents of our migrant population.

I'm not sure how often my father called or came to visit us in Collaroy. But within a few months my parents were reunited and we were all living together again. In 1951 they bought a house in Wollongong with another refugee family; we lived in one half of the house, and the other family lived in the other half.

We kept everything pretty separate in my memory, except for the toilet, which was in the backyard. The dunny men would empty it each week. Nobody had automatic washing machines. I remember my mother lighting a fire with sticks under a huge copper pot every Monday morning to wash the clothes.

Eventually, my parents bought out the other family. It became our family home for many years, and it was a lucky house for us. We were happy, and we prospered. (I've since visited the house as part of the *Julia Zemiro's Home Delivery* series, and I met the family who moved in after my mother sold it to move in with me in Sydney in the 1980s. It's been a lucky house for that family as well.) My father made the front gate and had one of the Water Board welders sculpt my mother's name, Rina, in wrought iron, across the top.

The block of land was large, a standard old quarter acre. My father kept us fed with a huge vegetable patch that was about 10 by 15 metres, in the back half of the property. He knew nothing about gardening when he started, but he learned quickly and developed a deft green thumb. We ate the vegetables he grew all year round. He also raised chickens, both for eggs and for eating.

My father was a strange mix. Even though he philosophically remained very much anti-communist, many of his best friends were practising communists. Perhaps the friendships were coincidentally linked to them supplying him with all kinds of dubiously sourced items, like timber for building and coal for keeping us warm. It seemed that his comrades were living true to the old communist manifesto that property is theft!

I don't think my childhood was unhappy, but I was a bit of a loner. I had a dog I loved, a smart brown kelpie. One day a neighbourhood kid and I walked to the newly built

Piccadilly Arcade next to the Wollongong railway station to see this amazing thing called a multilevel car park. A car park alone was amazing enough (because nobody had cars), but a multilevel car park was *wow*!

We walked up the ramps until we got to the top level. Okay, that was only a third level, but in Wollongong in the 1950s, this counted as pretty exciting. Then a bird went flying past us at eye level, barely a metre away. My dog jumped over the car-park fence to try to chase it, not realising there was a drop of several metres to the ground. He landed with a loud thump.

We rushed down and he was just lying there, not moving. I was ever so worried, but after a few minutes he came good. He walked home with us slowly, and by the next day he was back to normal. Lucky it was the dog and not me who fell off the third floor!

Once my dog had safely recovered, I was very excited about the whole adventure. I even rang up the local newspaper, the *Illawarra Mercury*. (By then we had a phone in the house – very posh for the time.) The reporter dutifully wrote down the details, and the very next day there was an article about how a dog had jumped off the top of the brand-new Piccadilly Arcade car park and survived! That's what you call a good-news story in a small town.

Despite little windows of joy, I have plenty of memories from my school years of being lonely. I wasn't especially sporty, I was very skinny and I wore glasses. None of these were attributes that got you many friends. On top of that, I was a 'wog' from Central Europe, or a 'New Australian', as we were called. It made no difference that I had been born in Scandinavia. Anyone who wasn't born in Australia or the United Kingdom in those days was called a wog!

I realised that I was 'different' in the early 1950s when the east coast of New South Wales was hammered by weeks of constant and very heavy rain. Back then, kids made their own way to school, usually on foot, and very few people had a car. But during that exceptionally heavy rain, the other kids in my street got picked up and driven to school in a car belonging to a wealthier family. But no one ever offered me a lift as I squelched along the path to school in my plastic rain hat and coat. The other kids would wave cheerily to me as they sailed past in dry comfort. And yes, I do think I was left out of the rideshare because I was a 'New Australian'.

There is one tremendously scarring childhood experience that I've never forgotten. I was out shopping with my parents and we were speaking to each other in Polish when a kid from my Catholic primary school class was pushed forward by his parents to declare, 'Speak English, you bloody wogs!' I was so shocked that he would speak to my parents like that; for me as a small child, it was a source of deep shame. I can't remember how my parents reacted.

In a knee-jerk reaction, I decided there and then to never speak anything but English so that no one would ever say anything like that about me again. Of course, this was a stupid decision, and I totally missed the point that no one should be shamed for being able to speak more than one language! As an adult, I see all that I lost by not remaining fluent in my parents' first language, Polish. But as a child, I was so scared to not belong.

Because of this decision, I effectively removed myself from my parents' conversations (and any ability to successfully order food in Polish restaurants in the future!). At mealtimes, I would read while my parents talked to each other in Polish. It must have been awful for them to have me cut them out that way, but when I was young, I couldn't see things from their perspective.

On top of my father's paid work at the Water Board, he unofficially provided assistance to fellow immigrants with all sorts of documents. He made use of his past legal training to help them fill out tax forms and licences or apply for passports. He was terrific at paperwork and was usually paid in whatever goods – eggs, timber, coal – that his fellow immigrants brought to show their appreciation.

My father's first job in Australia was working as the billy boy on a Water Board sewer-pipe gang of ten. Their job was to manually dig trenches to install the pipes. As billy boy my father looked after the hut and worksite, kept the tools in good condition and provided tea for lunch and morning and afternoon smoko. There were no gas bottles in those days, so he had to keep a wood fire burning.

My father made his way further up the ladder and was promoted to head office in downtown Wollongong in the personnel department (now human resources, or just HR). His ability to speak twelve languages was an obvious asset as he could talk with potential employees in their native languages, which must have made life easier for some of his fellow migrants. One day a man came to apply for a job, and my father immediately picked up on something distressingly familiar: he recognised him as a particularly brutal Nazi concentration camp guard. My father had personally witnessed this man beat and kick a prisoner to death – and also kill another inmate, but in a more subtle way.

The prisoners were forced to do regular exercise inside the camp, either to test German footwear or inexplicably and ironically to keep them in 'good health' while being simultaneously on a pretty much guaranteed pathway to death. One day, in the middle of the compulsory walk, one of my father's companions just fell to the ground. He sat there and refused to walk any further, sick of the hopeless situation: if they were all going to be killed, he couldn't see why he needed to wait any longer. My father and his friends tried to make him stand up again, afraid that the guards would simply beat him to death or set the dogs on him.

This same guard, now sitting in a chair in front of my father in the Wollongong Water Board personnel office, had commanded the other inmates to leave the prisoner on the ground. As the other prisoners completed each circuit, they begged him to stand up, but all the fight had gone out of him – he just wanted to die. He stayed sitting there, and to the surprise of the other prisoners, the guards left him alone. Night fell, and in the morning he was dead, even though it was only a very mild night. He did not have a mark upon him – he had just given up on living, so he 'willed' himself to death. This prisoner's death really broke the spirit of some of the other inmates, crystallising how little control they had over anything. Their own deaths in this miserable prison were very much guaranteed to happen – the only unknown was 'when'.

And now this cruel man was in Wollongong, applying for a job on the Water Board. My dad went through the usual administration in English – all business. He filled out the forms and finished the paperwork. Not even the merest flicker of expression on his face revealed what he knew. He said in English, 'We're very happy to have you on board. You can start next Monday if you'd like.' He paused, leaving a significant silence hanging in the air before switching to fluent German. 'I remember you. I know you from when you were a guard at Sachsenhausen concentration camp. I was a prisoner. I saw you then, and I see you now. You know what I am saying is true. I saw the terrible things you did.' And then my father said the guard's actual name – which was different to the name on the job application.

The man went white as a ghost. His life as a brutal prison guard must have flashed before his eyes, and memories of his past and the war crimes that he had committed flooded back. This totally unexpected conversation was threatening his fake identity on the other side of the equator and the world. This man was now totally dependent on my father's largesse and whim. The tables had turned.

The guard slumped in his chair, speechless. Eventually he stammered, 'What happens now?'

When my father told me his reply, I was shocked. He had every right to hate this man for his cruelty and to expose him as a war criminal. But he said, 'Nothing will happen. It will be our secret. I'll never tell anybody, and you'll never tell anybody.'

The former guard was confused, and after a long pause he asked, 'Why are you doing this?'

My father said, 'It's not for you. I'm doing this for your children and your family.'

That my father was willing to lock away the past and start afresh was mind-blowing to me. He put aside any personal sense of vengeance and left it behind. He drew a metaphorical line in the sand. I think that mindset is what allowed him to be able to mostly live his life without being haunted by past trauma. He was a forgiving man, and always hopeful that people could change for the better if you gave them half a chance.

I didn't know any of this history at the time it took place. All I knew was that at my primary school in West Wollongong, St Therese Little Flower, a new European child had just joined my class. This boy, roughly my height and age, was the child of that ex-guard. Sharing a migrant background, we became quite close, and my father never stood in the way of that friendship. We played together both in and out of school. We both liked reading library books – which not many of the other students did. On one occasion I even asked if I could take the bus to play at his home in another suburb, and my father agreed, despite what he knew. I went through primary school and high school with this good friend. He also studied at Wollongong University, and later went on to have a family and a very good career. But almost certainly none of that would have happened if my father had exposed his father for his war crimes. My father could see that nothing positive would have happened for the rest of the family if the father was imprisoned, even though he deserved to be in jail. Dad's compassion and optimistic hope that change would come for the better had a great influence on me, and my outlook on life.

My mother, however, was not the 'forgive and forget' type. I generally know so much less about my mother's life and thoughts and feelings before Australia, because she never felt truly safe to share her secrets.

One thing I do know is that, while in Sweden, she received an offer to work with Dior in Paris as a milliner. Perhaps she learned the rag trade while in the Łódź ghetto? Certainly she had a flair for clothing. She said she didn't take the Dior apprenticeship because it would have meant leaving her family behind. In Australia she placed an advertisement in the local newspaper offering the millinery services of 'Madame Rina, from Paris, Prague and Milan' – not a lot of truth in advertising in those locations! I guess it sounds more alluring than 'Łódź ghetto, Auschwitz concentration camp and Swedish refugee camp'. I still have some of her wooden 'hat heads'; I've polished a few up and use them as stands for the hats that protect us from the harsh Australian sun.

While I was in early high school, Mum was approached to manage a new record shop in the migrant suburb of Warrawong, which was directly under the pollution fallout from the steelworks. She built the business up from scratch and chose the name La Paloma for the store. The owner trusted my mother's judgement. He was very Anglo, so maybe he was trying to bring some European flair to the store by letting her pick this beautiful Spanish name. Being able to speak half a dozen languages helped with selling records to the multicultural community. Mum even designed and had built a special soundproof booth for customers to 'try before they buy'. She adored classical music but stocked anything that would sell, including The Beatles and Jimi Hendrix. For her, music was a source of culture and a connection to Europe.

During school holidays I would help in the shop. It was a fun part-time job, and I was always finding little chores that suited me perfectly. The records were sold in large paper sleeves. Quite separately, the record company reps gave us expensive full-colour promotional pamphlets for new releases, which usually sat on the counter and eventually got thrown out. I started inserting those brochures inside the record sleeves of each record we sold. I was thrilled when my mum told me that people were coming back and asking for a particular piece of music they had seen in the pamphlets. (I've never included this Marketing Prowess in my CV before this.)

I listened to whatever music happened to be playing in the shop. There was such a huge range. One Saturday morning I heard a Bach *Brandenburg Concerto* for the first time. I was enraptured, so we took it home. That afternoon I played the record over and over – I couldn't stop, louder and louder each time. It was transcendental! I fell in love. It was the first time I had ever experienced the power of classical music to tear at my heartstrings and totally consume me. I still play those Bach pieces on repeat, and I don't think it will be possible for me to ever tire of them. It was my first and most powerful experience of an 'earworm'.

Earworms

An 'earworm' is a piece of music that keeps repeating in your brain.

Sing the intro from 'Bad Romance' by Lady Gaga (the bit that starts 'Rah rah …') out loud. Odds are you're now going to have some trouble getting the verse out of your head.

One of the earliest uses of the word was in 1978, in the novel *Flyaway* by Desmond Bagley: '… "ear worm", something that goes round and round in your head, and you can't get rid of it'.

In general, an earworm needs three characteristics. It has to be part of your culture, the beat is a bit faster than usual, and it needs unusual features (e.g. incongruous or unexpected changes in pitch).

There are two 'cures' to get an earworm out of your head. One is to play it on repeat, until it eventually (and hopefully) 'wears out'. The other is to chew gum! It appears that by a nice coincidence, the pathways that are used for replaying music in your memory are also used for programming repeated jaw movements. But if neither 'cure' works, go with the flow and enjoy the free music.

We never owned a car, but both my parents loved travelling. One unforgettable train trip was to Katoomba in the Blue Mountains in October 1957 when I was nine years old. After dinner, we gathered around the radio (yep, people used to listen to the radio together before and after TV arrived in Australia in 1956) to listen to the evening news. It was full of unbelievable stories about the Russians launching the first-ever satellite, Sputnik (which meant 'satellite' or 'travelling companion').

The first person known to see Sputnik move across the sky was probably Dexter Stegemeyer of Fairbanks, Alaska – while seated on his outside toilet! It was total chance. There's even a plaque mounted on the dunny to mark the significance of the sighting. The toilet story gets better. Obviously it's pretty cold deep in an Alaskan winter, and back then, people would head to the outhouse in many layers of clothing as well as carrying a pre-warmed toilet seat in hand. A brilliant example of "local knowledge"!

Back to our holiday. Nobody had ever tracked a satellite before, but scientists swung into action, and the radio and newspapers told us we should be able to see it with our naked eyes in the early evening of 9 October. We walked up the hill to Katoomba railway station; usually the streets were pretty empty in the evenings, but on this night they were crammed with people. Somebody cried out, 'There it is!' It looked as bright as one of the brightest stars, but, unlike a star, it moved rapidly and steadily across the sky, until it disappeared after three minutes. Stars are fairly bright, but they stay stationary in the sky relative to other stars. Meteors are also bright and move against a background of stars, but they are usually visible for only a few seconds.

What we saw that night on the streets of Katoomba was very different from anything humans had ever before witnessed in the night sky in the 200,000-year history of *Homo sapiens*. (No, I don't see any evidence for the claim that aliens have visited us. But I really wish they would!) Sputnik tracked deliberately and awesomely across the sky, taking a few minutes to do so. And that night, I became one of the first (along with millions of others around the world) to ever see a satellite.

> **Sputnik 1**
>
> Okay, I'll come clean – I probably did *not* see Sputnik 1, the highly polished 58-centimetre, 93.6-kilogram sphere. It reflected light well but was also quite small, so it was not very bright. What I almost certainly saw was the much bigger R-7 'core stage' that launched Sputnik into orbit and was about 26 metres long, weighed 7.5 tonnes, and closely followed the much smaller (and practically invisible to the naked eye) satellite.
>
> Astronomers measure the brightness of objects in the night sky in 'magnitudes'. The first astronomer to do this was Claudius Ptolemy (100–170 CE), who was also probably the first known person to actually count the

> number of stars visible on a cloudless
> and moonless night. For astronomers, the
> reference for brightness is magnitude 0,
> which is roughly the brightness of the star
> Vega, the brightest star in the northern
> constellation of Lyra.
>
> The core stage of the rocket that
> launched Sputnik was a magnitude 1
> brightness. At 40 per cent the brightness
> of Vega, it was easily visible.
> (Confusingly, the bigger the value of
> magnitude, the dimmer the object is.
> Sorry.) But the actual Sputnik 1 satellite
> only had a brightness of magnitude 6, about
> 0.4 per cent of the brightness of Vega, and
> was only barely visible to the naked eye.
> (The generally accepted lowest level of
> visibility is magnitude 6.5, about 0.25 per
> cent of the brightness of Vega.)
>
> Weighing up all the data, I guess on that
> memorable night, in that strange gathering
> around the Katoomba railway station, we
> likely saw the large body of the launching
> rocket, not the much smaller Sputnik 1.
>
> But it was still very memorable.

For my father, this Soviet satellite launch was shocking and intensely unsettling. Here he was, all the way on the other side of the world, trying to get away from the Soviets, but now they were flying over his backyard! He was uncharacteristically gloomy about the whole event, and deeply worried this meant the Soviets would be dropping atom bombs on us any day now.

Society was beginning to undergo big changes, and Sputnik was an early harbinger of that. I was only a primary school kid, but the gravitas of the night, and a fear of future Sputniks releasing atom bombs, planted the first seed of nuclear anxiety in me.

Which brings me to my time at school, the later years of which were played out in harmony with a musical backdrop of Bob Dylan's 'The Times They Are a-Changin''.

1954–64

School years

Babies are always precious, but in the aftermath of war and devastating loss of human life, where 2 per cent of everybody on Earth died, babies born after the Second World War were seen as extra precious. The label Baby Boomers, given to my generation was itself born from the significant bump-up in babies after the war.

    I was in primary and secondary school from 1954 to 1964. This made me part of the first generation of kids born and educated after the horrors of the Second World War. My parents sent me to local Catholic schools – six years at the coed St Therese Little Flower primary school, followed by five years at the boys-only Edmund Rice College. Most of the teachers were nuns at St Therese, or Christian Brothers at Edmund Rice. There were some 'lay' teachers around who were still Catholic but not part of any religious order. Surprisingly, none of the teachers 'teaching' in Catholic schools at that time were required to have any formal training or qualifications in education.

I remember on one Saturday morning, my parents and I walked into central Wollongong (we generally walked everywhere to save money) when my father spotted one of my primary school lay teachers. My father told me to do what he did. As she got close, my father stopped (and so did I) and bowed slightly to her while taking off his hat (and so did I). She slowed down, acknowledged us with a lesser bow, thanked us and kept on walking.

My father told me to always respect teachers, and I have.

Back then, in high school you received an Intermediate Certificate after third form (Year 9) and a Leaving Certificate if you stayed on to fifth form (Year 11). If you passed the Intermediate Certificate you were eligible to train for a trade, and passing the Leaving Certificate allowed you to enter a university.

I started high school when I was eleven and finished when I was sixteen. I was young because my March birthday bordered between one year's intake and the next, so they shoved me into the slightly higher class. That's how I entered university at the age of sixteen.

At my high school, the only subjects offered were science, maths and languages (English, French or Latin), so I just did what was offered. There was no history! And 'science' was very limited – no geology or biology, just physics and chemistry. I enjoyed learning and understanding new concepts, but the style of teaching in the classroom back then was more rote learning with harsh and slightly terrifying discipline. There was virtually no awe and wonder. But I got that from my extra-curricular science-fiction reading. I had absolutely no insight then that science would be my ultimate destiny!

Cold War

Following the elation at the end of the Second World War came the anxiety and fear about the Cold War between the West and the East. The difference between a 'cold war' and a 'hot war' is the lack of open hostilities (battles, bombs and bullets) and the much, much lower death rate.

The West saw communism as the universal enemy of all 'freedom-loving peoples'. The two superpowers – the USA (with its allies, including Australia) and the Soviet Union (with its allies, including China) – were pitted against each other in a deadly stand-off thanks to a nuclear-based doctrine known as mutually assured destruction (MAD). If the USA and the Soviets had ever launched all their stockpiled nuclear weapons at each other – and they each had about 25,000 nukes in the late 1970s – the result would have been total nuclear annihilation! This would have meant well over three-quarters of humanity dead within a year, no electricity, no schools, no hospitals, no factories, no antibiotics, no manufacturing – just utter destruction. Luckily that did not eventuate, but it was a big and very real threat to live under.

MAD

MAD was the acronym for the principle of military deterrence known as mutually assured destruction. The best image to understand the craziness of MAD is that of two old white men standing up to their necks in a swimming pool of petrol. One of them is holding five burning matches in one hand just above the surface of the petrol. The other has six burning matches and cries out triumphantly, 'I win!'

Of course, nobody wins and everybody loses if either of them drops a single match, and in the case of the Cold War, 'everybody' meant everybody on Planet Earth.

We have definite proof of three cases where we barely avoided total nuclear holocaust. In two cases, relatively junior American and Soviet officers (at separate times, and in different locations) received incorrect radar readings that suggested the 'other side' had launched a massive surprise nuclear missile offensive. In each of those cases, the junior officers broke protocol and did *not* pass the incorrect information up the chain of command. We don't really know why they disobeyed protocol. Perhaps it was a mixture of doubting the accuracy of the radar

A Periodic Tale

> information, or perhaps a gut feeling, or perhaps a desperate desire to avoid the escalation that would lead to the devastation of humanity. The third case, the Cuban Missile Crisis, happened in 1962, when I was in third form in high school (see page 44).
>
> Humans alive today owe our existence to those three people.

My father had been captured by the Soviets twice during the war and, incredibly, escaped each time. He always believed that his unfinished Russian jail time would likely come to rest on my head if I ever laid foot in the Soviet Union. It turns out that he had a valid reason!

After escaping the prison in Grodno with another prisoner, my father and his fellow escapee both somehow ended up in Wollongong. They stayed friends, and years later, when I was at the University of Wollongong, the son of my father's friend, against the strong advice of his family, went to Moscow during the Cold War on a holiday. At the end of his first day of sightseeing, he returned to his hotel to find a nasty surprise: his room was full of KGB operatives standing around a suitcase he'd never seen before! (The KGB was basically a rebranding of the Russian secret police, the NKVD, and operated from 1954 to 1991.) They opened the suitcase to reveal US$100,000 and he was arrested on the spot, and then quickly tried for smuggling currency. He was sentenced to the exact amount of time that was remaining on his father's sentence – clearly *not* a coincidence. My parents read it as a message from the Soviets that they were still watching anyone who had ever been an enemy of the state, and that potential revenge would be dealt out to not just my father, but the entire family.

During my time at school, my father continued his day job for the Water Board, but in his spare time ramped up his journalism. He worked for both the Australian and foreign press, as well as being the Australian correspondent for Radio Free Europe, then funded by the US Central Intelligence Agency (CIA). He mailed his voice recordings to Europe each week and joked that my school education was paid for by the CIA! Sometimes he would get me to put on a fake accent and pretend to be someone he was interviewing.

In 1962, when I was in third form at high school, the Cuban Missile Crisis erupted, a thirteen-day window of high tension between the Americans (who had installed nuclear-tipped missiles in Turkey and Italy, relatively close to Moscow) and the Soviets (who tried to counteract by installing nuclear-tipped missiles in Cuba, just 140 kilometres from the USA). Even though we were on the other side of the equator, my parents were very worried, and as a result, so was I.

We didn't realise at the time how close we all came to annihilation, because the real story only came out decades later. During the crisis, USS *Beale* attacked the Soviet submarine *B-59* with signalling depth charges to force it to the surface. The submarine in fact had a nuclear-tipped torpedo and was authorised to launch its nuke if attacked.

The Soviet captain, believing he was under fire and following official protocol, intended to launch his 10-kiloton nuclear weapon at the American fleet. (For comparison, Hiroshima was obliterated by a 15-kiloton nuke.) Luckily for the world and you and me, Vasily Arkhipov, chief of staff of the submarine flotilla, was unexpectedly on board. He personally overruled the launch command, which would surely have seen the MAD policy swing into action.

But at the time we lived in blissful ignorance. Back then, most news came over the radio or in newspapers. It was only in my teenage years that I very slowly picked up an awareness of the world outside of Australia – a 'world view'.

That world view started to expand with TV. Around 1958, I rode my red scooter about a kilometre in the dark to visit a wealthy family from my primary school who had both a car *and* a TV. I remember the thrill of watching *I Love Lucy* on their black-and-white screen. (Forgive me for linking *I Love Lucy* with a 'world view'.)

My school life might have been set against the ominous backdrop of the Cold War, but school itself was for me defined less by politics and more by obedience. Obedience was expected and enforced, prejudice was everywhere, religion was drilled into me, sport was something I missed out on, the library was a mind-changer, polio was rampant and change was coming. What a mixed bag!

Obedience

At school in the 1950s and '60s, absolute obedience was a given. You simply did what you were told, otherwise you got the cane. Discipline was strongly enforced by physical punishment, usually on the hand.

In primary school the nuns used a thin cane, while in high school the brothers had a leather strap. It was about 30 centimetres long and square in cross-section, about 3 by 3 centimetres. It was simple: you put your hand out and the teacher would hit it as hard as they could, and as often as they felt like. My parents had never hit me, so school was my first exposure to corporal punishment, and I didn't want a bar of it. I was a good student and usually followed the rules, which mostly saved me from being punished.

But others did get hit a lot. In the 1962 Edmund Rice year book, the principal wrote, 'Most boys are amenable to discipline. Occasionally we encounter a pocket of resistance but these usually conform after suitable remedial measures have been taken.' Suitable remedial measures? I can't accept that physical violence in schools is ever allowable, much less desirable.

On one occasion in third form, I went back to my classroom during break because I had forgotten to get my food from my bag. I knew it was forbidden to go into the classroom during lunch, but I was hungry, which meant I was prepared on this occasion to break a rule. As I passed the back of the classroom, I was terrified to see one of the brothers, a short blocky man who used to be a boxer, punching a student in a very methodical way that I now suspect was intended to leave no bruises. I backed away immediately without my lunch and much to my regret never mentioned this to anyone – not my fellow students, not my parents. As to punching as a 'suitable measure', I don't think so!

The principal also wrote that 'one of the sources of poor behaviour and moral decline in some boys is the Surf Club and sometimes the football club'. He failed to mention, of course, the sexual predation of some of the brothers upon my fellow students. In 2013, the Royal Commission into Institutional Responses to Child Sexual Abuse was finally established after decades of allegations going back to the 1950s. It found that many Catholic children had been sexually abused by their teachers and priests, and that the perpetrators had been moved to new locations after the allegations. And it was all part of a deliberate cover-up by the Catholic Church.

I remember some of my high-school teachers (both Christian Brothers and lay) openly cuddling students in the front of the class. I don't know if this public physical contact was followed privately by sexual abuse, but it was strange in a school where usually the only public touching was with a cane. I didn't give it any thought at the time and don't remember having any knowledge or awareness of sexual abuse as a child. I know now at least one of the Edmund Rice students died by suicide a few years after leaving school. A few decades later, I met up with one of my fellow students who told me that paedophilia had been widespread. He told me that at least two of the teachers in my five years of schooling had sexually assaulted several of my fellow students. I realised then that I was fortunate to have been left alone.

I paid attention in class and dutifully did all my homework. I was also lucky that my parents supported me in my studies and wanted me to do well. Not all parents thought that their kids needed to excel at school – but my parents encouraged me to try hard. Of course, good school marks are not the only way to set you up for success in later life – but sometimes it gives you more choices. I was usually near the top of the class – however, the school wasn't particularly strong academically, so it wasn't a huge achievement on my part. Again from the 1962 year book: 'our examination results were good … 25% of the Leaving Class failed but I can assure you that they failed despite the tremendous efforts of a very zealous and much experienced teaching staff … Some boys failed because they resisted all the efforts of parents and teachers to make them work. This type [of student] is always with us …' That was awfully mean-spirited, as well as just plain wrong. Back then, there was very much a culture of blame when it came to students. It seems shocking and heartless that the principal could excuse the staff of any responsibility when one quarter of the students failed.

Astonishingly, the principal made a simple mathematical error! He wrote, 'Thirty-six boys sat for the Leaving Certificate, and twenty-nine passed.' That's 19.4 per cent that failed, not 25 per cent. Maybe his inability to do basic maths reflected the standard of the teaching? It's a wonder more students didn't fail, if even the teachers couldn't do their sums.

Prejudice

Prejudice was thoroughly embedded in Australian society during my school years and across very many different layers. Women suffered much discrimination. Sure, they were now allowed to vote, but they were paid less than men (and still are!) and in most cases had to give up their jobs if they got married. Aboriginal people were subject to dreadful racism on a political and personal level. Back then it was still official government policy to kidnap Indigenous children from their parents and send them away to become unpaid or cheap labour, resulting in the Stolen Generations. Migrants from practically anywhere outside of Australia were also the subject of prejudice. And just to keep the hate flowing (haters gonna hate), Catholics hated Protestants and vice versa. Basically, on every level there was someone who hated another group for no other reason except that they were different – and back then, it was generally unchallenged.

Immediately after the Second World War, virtually everybody in my hometown of Wollongong, about 80 kilometres south of Sydney, came from an Anglo-Saxon background. All the way through my schooling, my fellow students were overwhelmingly white and Anglo-Saxon.

We migrants were different. I had more friends at high school, but almost none were the Anglo kids. My friends were mostly the children of refugees from Europe, and we were very much the minority group. Over my last three years of high school, about once a week, somebody would let down the tyres on my bicycle. I never found out who did it or why. On a lighter note, I also remember having black bread for my sandwiches at school while practically all my fellow students had soft white-bread sandwiches.

One other memory of still being an 'outsider' stands out. It relates to the Catholic priest. He was surprisingly wealthy, and had a convertible Cadillac that he took the church altar boys for rides in. One day we got new neighbours from a very Irish background, and the priest came around to welcome them the very next day. At dinner that night I must have said something about the priest's visit, and my father shot back that he would never visit us, because even though we were Catholic, we were not Anglo-Saxon. Then my dad went further. He said he didn't like the priest's sermons either, because every single sermon asked for more donations to the church. The priest never discussed anything of a moral or ethical nature. My father was a very religious man who never usually spoke against the Church, so it really shocked me to hear him criticising the priest so directly.

Religion

Religion was simply there all the time. I went to Catholic schools, and at home my father followed strict Catholic beliefs. For me, church was incredibly boring and uninspiring, but I went with the flow and just thought sci-fi thoughts.

In my final years of high school, students were pressured to commit to entering a religious order as a priest or brother. As part of the push there was a weekend retreat to immerse us in religiosity in the big smoke, Sydney. Our time at the retreat was filled with a mix of Mass, Stations of the Cross, confession, praying in groups and silent retreat, where we were supposed to think deeply. Somewhere over the weekend, each student had an allocated interview. Mine was with a Jesuit priest, and I knew the Jesuits had the reputation of being the 'intellectual' branch of the Catholic Church, so

I was very excited to ask my big question: 'Can you give me the proof that God exists?'

He looked at me carefully, and I assumed he was basically going to repeat what I had already been fed the previous year in my religion class: everything in the universe has a cause, therefore, to start the whole universe going, there has to be an original 'uncaused cause', which today we call 'God'. This concept was documented a few thousand years ago by Aristotle, who called it 'the unmoved mover' or 'that which moves without being moved'. The Catholic theologian Thomas Aquinas expanded and elaborated on this concept in the 1200s, but I could see holes in that unsophisticated logic. I didn't say this to the Jesuit, though – I specifically asked only about a proof for the existence of God. Surprisingly, the Jesuit priest said to me, 'The "uncaused cause" argument is full of holes. Can I absolutely prove to you the existence of God? No. You've got to have faith. There is no proof for the existence of God.'

In that moment, any religious belief that I had faded away. I settled into a position where I basically still sit: an agnostic. (From the Greek, 'a' means 'not', and 'gnostic' means 'to know', so an agnostic is someone who says they do not know.) I don't know if God does or doesn't exist, and I'm not absolute enough to be an atheist who believes there is no such thing as God.

I think religion and science are both valid but work on totally different assumptions. Science needs proof and has no room for faith, while religion runs on faith and proof is irrelevant. Both science and religion can happily exist in the same person's brain without conflict. At the University of Sydney, I met a professor of physics who was simultaneously a deep expert on Einstein's theory of relativity and a minister of religion. He saw no conflict and happily took from both.

I tend to follow Einstein's belief system: God set up (or is) our universe – and doesn't care about us as individuals.

After all, several million children under the age of five die each year due to a lack of food, medical care, vaccinations and so on. They did nothing bad, apart from being born into poverty. Why doesn't a loving God fix this?

So I am most happy to be a member of the Church of God the Utterly Indifferent, who set up the universe, but is now off playing golf (or something). (I found this wonderful concept in Kurt Vonnegut's sci-fi novel *The Sirens of Titan*.)

Sport

Rugby league football was the only sport offered at my high school. I wasn't built for footy. I was slight and wore glasses, and rugby league was brutal. The other students deliberately ran into me even when I wasn't holding the ball, and if I *was* holding the ball and got tackled to the ground, they would punch me. No wonder we migrant kids nicknamed it 'thugby'.

If you didn't play footy you could always 'choose' to pick up rubbish in the schoolyard. Yep, that was my 'sport' for my entire five years at high school. It's not that I didn't like sport – I just didn't like getting beaten up. But it wasn't all bad – that early training stayed with me. I'm the bloke who picks up other people's rubbish whenever he goes for a walk.

It turned out that I was a natural at tennis. We could never afford lessons, and my playing style is still decidedly odd, but it gave me a real thrill to hit a ball over the net. You had to wear all whites to play, with the only colour allowed being stripes around the neck of your jumper. All the racquets were wooden. The only difference between them were the strings: nylon ones were waterproof but a bit stretchy, and cat gut ones were less stretchy and gave more control but were also more likely to break in the wet. (By the way, cat gut comes from cows, never cats!)

When I was about twelve years old, I had a strange and unexpected experience during a tennis match that changed me forever – it was the death of my 'competitive spirit'.

I have worked with very competitive colleagues who are continuously driven to be number one. I don't get it. For me, competitiveness has nothing to do with other people. I have my own standard, and I try to lift that standard. Mind you, I am very happy to learn from other people. But I'm never driven by a desire to beat them or be better than them.

Survival of the fittest

```
In evolution, survival of the fittest
has nothing to do with big muscles. It
refers to natural selection, which is a
type of competition. In the short term,
fittest means having more babies than your
competitors. In the long term, fittest
```

> means their descendants die out – but
> yours don't. Through the lottery of
> sexual reproduction you get different
> characteristics, so the surviving offspring
> might be able to make a life in new or
> different conditions.
>   Evolution doesn't have to be perfect,
> just good enough to have more babies.

This life-changing cosmic flash about competitiveness happened after a tennis match. (Oddly, this story about not being competitive was born out of playing a very competitive-sounding game – cutthroat!) There were two brothers at my school with whom I played tennis most weekends on the free school tennis courts. In our version of cutthroat tennis, two people played on one side of the net using the doubles court and one person played on the other side in the singles court. After each game you rotate your position, usually clockwise, so for every third game you're on the singles side, playing by yourself.

Back then I was a terrible cheater – at least in tennis. When I was on the singles side, I made line calls that were deliberately in my favour. My excuse was that I could see the ball better, because I was so much closer. But the strange thing was that the brothers didn't seem to care if I called the ball in or out. They just kept on playing, accepting my ridiculous line calls. At the time I never really thought about why but simply took the points. Sometimes I 'won' the set, or match, only because I cheated.

One day, after I had cycled home in the afternoon as usual after a match, my parents asked me how the game was. Something (and I still don't know what) was different in me, and for the first time I listened and responded to the actual words: 'How was the game?'

I suddenly realised the game was terrific and I'd had a really good time. It didn't matter whether the score was six games to me and four games to them or the other way round. The fundamental thing was being with friends and having fun. At that exact moment, as I really listened to what my parents asked, I lost my competitive spirit in the sense of trying to beat another person rather than beating myself. It was a turning point, and I never cheated in tennis again.

A few years later I won the 'highly sought' Under-16s, Division 2, Milo Cup Tennis Trophy for Wollongong. I know it's not Wimbledon (so lucky I'm not competitive anymore), but it has a special spot on my trophy shelf. And that's because I love the game.

Back then, I could easily play five sets in a match without getting puffed, even on hot days. I drank water all the time when I was exercising and never got a stitch, but I have a clear memory of being told not to drink water because it *would* give me a stitch. Odd!

Playing sport at a high cardiovascular level for a decade or so in one's teenage years is good for long-term health, but I only got to do that on weekends. I wish that I could have played more sport in school hours. (I am, however, still pleased at how good I am at picking up rubbish.)

Library

I love reading and have since I was very young. I found solace in books.

By the time I could read fluently in early primary school, my father had been promoted from labouring to the Wollongong head office of the Water Board, and that building was right next to the library. So after school I would take the bus to the library, borrow half a dozen books, and then travel home with my dad when he knocked off. Sometimes, he let me press the button for the 5 pm finish-work siren.

The Wollongong library really looked after me. At one stage I became fascinated by fairy tales and started reading everything I could find. The librarians noticed – and ordered in a complete set of fairy tales from the 200 or so countries on Earth, just for me! I soon became astonished at just how many similarities there were in fairy tales across the world.

It was a very natural progression from fairy tales to sci-fi. I still remember the very first sci-fi book I read: *Thunderbolt of the Spaceways: The Story of a Daring Pioneer of the Twenty-second Century*. It was published in 1954, and I read it soon after that – mind-blowing. I then moved on to the Kemlo children's series. The heroes were kids just a little older than me who were born on a space station called Satellite Belt K. They all had names beginning with K (just like one of my names!) and could live and breathe in the vacuum of space without any assistance. OMG, my head spun, and in the mid-1950s, I went to sleep dreaming of being one of the K-kids in space!

The librarians of my childhood weren't the grumpy ones in movies who always shushed. They went out of their way to cater to, and nurture, my natural inclinations, always looking at what I was reading and making suggestions. I profusely thank those generous librarians, whoever and wherever they are.

Libraries are the light in the darkness!

Polio

One very strange (but totally true) memory is getting beaten up by a kid with polio.

Poliomyelitis is a highly infectious virus that has been described and depicted for thousands of years. It can paralyse muscles – especially leg muscles, and usually just on one side of the body – by attacking and destroying cells in the central nervous system. Without instructions from the central nervous system, muscles can't be activated. In the short term, you lose control of those affected muscles. In the long term, that group of muscles wastes away from lack of use.

When I was in primary school, polio was a very common and much-feared disease. Walking around Wollongong, you'd see somebody who'd had polio at least once a day using either walking sticks or supportive 'leg irons' (also called callipers) to get around. Several of my fellow primary school students wore leg irons.

In the early 1950s, polio was rampant across the world until the early polio vaccines became available mid-decade. One day, without any warning, all of us students at the

Little Flower got loaded onto buses and driven to a hall in downtown Wollongong. As I walked into the hall, I saw my parents waving to me as they stood with the other parents along the walls. What was going on?

Then, one at a time, we students were injected with the first doses of the Salk polio vaccine available in Wollongong. It wasn't the best vaccine, because it didn't give 100 per cent protection, but it was so much better than no vaccine at all.

The adults waited till more vaccine doses were available. I guess it was recognised that the kids had the most to lose in terms of how long they would have to live with the disease, if they got infected.

Decades later, when I was a medical doctor and understood polio a bit better, I realised that our society had done a wonderful thing back then. That kind of behaviour for the common good makes me proud to be a human.

A kid in my street had mild polio. He was quite strong and got around reasonably well, but he did need crutches. Whenever he saw me he shouted out, 'Go home, ya wog!' My father told me not to be scared and to stand up to him. (Unfortunately, my father did not follow through with any boxing or martial arts training!) So the next time the boy launched into his regular insults, I stepped up close and said, 'I'm not scared of you.' (That was pretty much my whole plan, right there.) He lunged forward, punched me really hard in the nose, then really hard in the stomach and I fell down. While I was lying on the ground, he hit me many times with his crutches. So yep, I got beaten up by a kid with polio.

It sounds like a comic book story now, but it was awful to be hit that way as a kid.

Years later, in my very junior medical years, I was in training with a GP in northern New South Wales. I was (and still am!) super impressed by how much knowledge a GP needs every day to deal with so many varied patient presentations. One day, two parents came in with their sixteen-year-old. As soon as the teenager had taken three steps, I recognised the absolutely characteristic 'polio walk'. So did the GP. We looked at each other and he widened his eyes, which I took as a message to say nothing.

The doctor was very open and gentle, asking the history. The family had been to Asia for a holiday, where the teenager had come down with a bit of 'food poisoning' and then developed a slight limp with some leg weakness. The food poisoning went away, but the limp and leg weakness persisted. Interestingly, the parents did not suffer any 'food poisoning', even though they ate the same food.

The GP asked some more general medical history and finished off asking about their kid's vaccination history. At this stage, the parents ramped up immediately and became very angry, saying that 'all vaccines are poisons' so their child was never vaccinated. When they asked the GP what he thought the problem was, he said polio was definitely high on the list.

Suddenly the parents went wild and totally lost it. I was so unprepared for their passionate response. They stopped being logical. They were shouting that polio was not a disease that harmed you, and also that polio didn't even exist. But somehow, despite the contradiction, if it had existed, polio had already gone away due to better living conditions, not the vaccines. They repeated over and over that all vaccines are poisons. In the end, after throwing furniture around the GP's office and breaking a chair, they stormed off, their teenager limping behind them. Before then, I didn't even realise that being 'anti-vax' was such a divisive thing.

It turned out that the teenager did have polio, and the symptoms that the parents took for 'food poisoning' were almost certainly from the initial infection with the polio virus. The GP empathised that perhaps the parents were so angry because a subconscious part of them realised their beloved child was sick with a totally preventable illness that they had chosen not to vaccinate against. And why didn't the parents get polio? Because they (like me) had been vaccinated as children.

Change

Change was in the air (just like love) during my high-school years. And I don't just mean rock 'n' roll!

In my last year of high school, our local university got hold of a newfangled machine called a 'computer', and they let final-year maths students from some of the schools have a play with it. The computer, an IBM 1620, was huge and blue, with its accompanying tape drives, hard drives, card readers, printers and air-conditioning units completely filling its own large room. The hard drive alone weighed over a tonne, and was the size of a big double fridge.

> **Big Blue**
>
> ```
> The International Business Machines
> Corporation (IBM) was nicknamed Big Blue
> for several decades. It was one of the
> first companies to get into computers, and
> for a long while IBM dominated this field.
> Early on, its psychologists worked out that
> people were a little scared of this strange
> new technology and reckoned that baby blue
> was the most calming colour. So for several
> decades, practically all IBM computers were
> baby blue.
> ```

The IBM 1620 was the first computer reliable enough to control factory processes in real time. It ran on the computer language called FORTRAN, and about 2000 of the machines were made between 1959 and 1970.

I had already read Isaac Asimov's sci-fi stories about artificial intelligence and his 'Three Laws of Robotics', and this early computer access gave me a hint of an 'intelligence' of some sort inside those big blue boxes. It was fascinating to 'talk' with this machine. It wasn't that smart, but it was smarter than any other machine I had seen.

I wrote my first computer program on this IBM when I was just sweet sixteen. Us schoolkids coded our FORTRAN programs on 'punched cards', a piece of stiff paper about the size of your open hand. The holes were punched into a grid of twelve horizontal rows and eighty vertical columns. It was so cool to be writing actual computer programs that simulated the movement of cargo ships in and out of the nearby steelworks port. This exposure gave me a tiny glimpse of the kind of change that computing would bring.

Music was also blasting out big changes. Rock 'n' roll and Elvis had stormed the scene – the kids loved it and the parents hated it. This first Beatles song I heard was 'She Loves You', and I just loved it to pieces. It climbed up the hit parade and then began to drop down. But unexpectedly, it then rose up the charts again – all the way to number one! This oscillation up and down and up again had apparently never happened before. My father thought it was a stupid song and would mock it, which began a bit of a split between my dad and me.

Internationally, things were changing too. In my last year of high school, the infamous Gulf of Tonkin incident took place, leading directly to an escalation of the Vietnam War, in which three million Vietnamese died. The United States falsely claimed that on 4 August 1964, North Vietnamese ships attacked US ships in the Gulf of Tonkin. This was used as justification for the escalation of US troop numbers, and then for Australian troops to land in Vietnam. The so-called attack never ever happened. That wasn't the first lie a government had told, and it definitely was not the last – remember the weapons of mass destruction that didn't exist and the Iraq War?

Carrots and night vision

On a lighter note, the British government was busy lying about carrots during the Second World War.

> They claimed the reason that British fighter pilots were able to successfully shoot down so many Nazi bombers during the Battle of Britain was because carrots gave their pilots better night vision. The propaganda was put out there to cover up their groundbreaking invention of radar small enough to be carried on fighter planes. Well-meaning farmers, believing the lie, would turn up at the local airstrip with bags of carrots to help the fighter pilots!
>
> And the myth still persists today.

My last year of high school was 1964. I didn't know at the time that the Gulf of Tonkin incident was a lie. I just knew the Vietnam War was ramping up. Supposedly it was going to save Australia from the red tide of communism rolling 'downhill' from Vietnam in the north all the way to Australia at the bottom of the world. People were genuinely fearful!

Despite the regional insecurity, I had personal surety about my immediate plans. On finishing school, I had a holiday job lined up as a labourer, and then I would enrol in university. I was excited to be finishing school and a little bit nervous. I had no idea what would come after I graduated!

1964–65

Digging ditches

My father had organised a labouring job for me on a Water Board gang. We were digging ditches as part of bringing a proper sewer system to Wollongong.

Working as a labourer is a fantastic way to get fit! I spent three months digging ditches. One of the lessons I learned, while working on the tools, was to always do the best job you can, even if that job was helping Wollongong transition from dunny cans to flushing toilets.

The dunny or outhouse was a small backyard shed built around the toilet, which itself was just a round metal can about 50 centimetres across and 60 centimetres high with a wooden seat and lid. Once a week, the dunny men came to empty the dunny can, full of urine and faeces, into the dunny truck before returning it empty to the outhouse. Everything about the dunny was stinky! You had to breathe through your mouth when you went, to minimise the disgusting smell.

A Periodic Tale

The boss of my Water Board gang was roughly the same age as my father and fairly strict, expecting all of us to work hard. We had one shed for our shovels, spades, picks, steel poles, woodworking tools, concrete-laying tools and more. The other shed was for our morning tea, lunch and afternoon tea – and shelter if it rained.

I turned up at 7.20 am on the Monday morning after my final high-school exams. I looked at the partly dug trench – about a metre wide and three deep – and realised it was constructed the same way the Romans built ditches two thousand years ago. As part of studying Latin, we had learned some of the military technology used by the Roman legions. I felt like an archaeologist, and I loved that sense of connection with history.

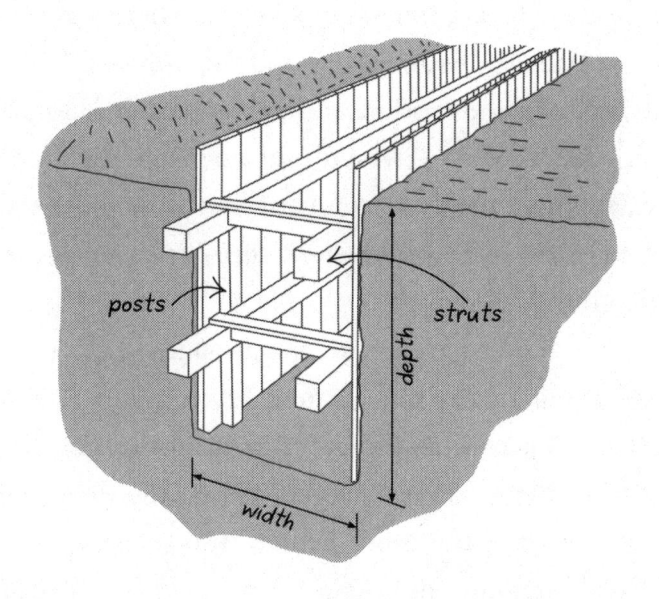

There were vertical planks pressed against the dirt on each side to stop it collapsing. Then there were crossbeams, or struts, to keep the vertical planks apart. A temporary wooden platform sat at a depth of 1.5 metres (it was too hard to toss up a shovelful of dirt the whole 3 metres, so this was a halfway point). The job was split in three parts: one labourer working at the bottom would toss the dirt up to the 1.5-metre wooden platform, another labourer would toss it up to the surface, then yet another labourer would toss it into a pile.

I was a naïve sixteen-year-old straight out of school, and I didn't even know how to use a shovel. The other labourers taught me how to plunge the tip of the shovel into the ground with my arms and then place the sole of my boot on the metal base to drive the shovel further. They taught me when to use a shovel (round head) or a spade (flat head), and when to choose the long or the short handle, or the one with the hand grip at the end. I realised quickly that there was no such thing as 'unskilled' labour – every job requires skills!

With practice, the motion of digging became smooth and efficient. After a while I would find myself transported into The Zone, where my entire universe was the shovel and the ditch. I loved it. We would work really hard in bursts and then swap over.

Once we dug the trench to the proper depth, the next step was laying down the ceramic sewer pipes – roughly 20 centimetres across, and about a metre long. The 'fall' or downward gradient of the pipe was critical: too steep and the liquids would leave the solids behind, too shallow and both the solids and liquids would grind to a halt. A bed of sand or concrete was laid down, the sewer pipes were joined together, and then the dirt would be filled back in. There was definitely an art to it! Everybody in the gang was proud of their work and did a good job.

I learned three lessons from my time digging ditches.

First, sewage travels inside sewerage. The trick I use to remember this is that sewage (smaller word) is the smaller, smelly solids and liquids that travel inside the sewerage (bigger word) pipes.

The second lesson is that without plumbers, we'd have no civilisation. Keeping sewage separate from our drinking water is essential for public health.

The S-bend toilet

I still reckon that the flushing S-bend toilet is one of the great inventions of all time. The oldest flushing toilets we know of go back about 6000 years ago to Mesopotamia, where clay pipes carried sewage away.

Queen Elizabeth I of England had a flushing toilet installed around 1596. But the problem with toilets of that era was the stinky smells of the sewer, which came up into the bathroom. The seal on the toilet lid was never good enough to stop the smells.

This was resolved by the amazing S-bend. First, the S-bend traps clean water in the bowl that catches and temporarily holds the urine and the faeces. Second, that water stops the smells from the sewer system from coming into the bathroom. Third, the seal (plain old water) never wears out and gets renewed with each flush. The water is the 'seal'. Wow!

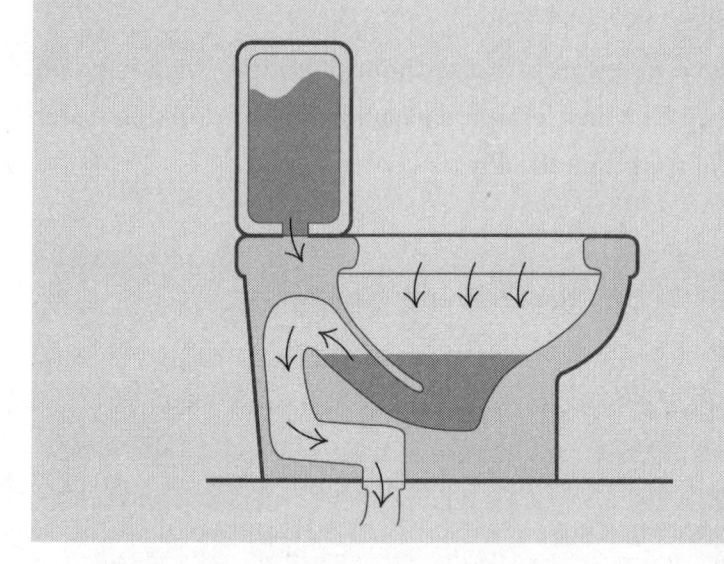

The third lesson was more personal and life-changing. It came from a fellow labourer who was in his fifties.

He called me 'Prof' like the rest of the gang, because they all knew I was just there for the summer before I went to uni. He said, 'Look, Prof, you're smarter than me and you'll be able to do anything you want in life. Me, all I can ever be is a labourer. A lot of people are like me. They don't have a lot of choice. I want you to never forget that.'

I had never really thought about that before. Smarts get handed out randomly – just like a lot of other skills – and you don't get any say in the matter.

I've never forgotten his words. Even though I had been taught about the concepts of compassion and understanding in my religious schools, it was all 'theoretical'. He was the one who really brought it home to me. He also made me think about the need to provide everyone in our society with an income they can live on. Random luck shouldn't mean you can't make a decent wage.

1965–67

Coffee, politics and uni – in that order!

I knew, because it had been drilled into me, that after school I was going to university. This was a very typical expectation for the children of migrants. Wollongong didn't have its own university, but it did have a campus of the University of New South Wales. I studied there from 1965 to 1967.

I did well enough in the school Leaving Certificate to choose any course I wanted. I also won a Commonwealth Scholarship, which covered all my university fees. I was sixteen, going on seventeen, and I had no idea what I should study.

I blithely (and foolishly) ignored advice from an old family friend, the medical doctor who had let us live in his big house in Collaroy all those years before. Our families had kept in contact after we moved out of his house into our own home in Wollongong. He would drive down Mount Ousley every Easter in his old Rolls Royce to visit us. When

I asked him why he didn't get a new one, he chuckled, 'If I get a new Roller, it shows I am wealthy now. But if I keep driving my old Roller, it tells people I was wealthy back then!' Our family didn't even have a car at that time, but I still found it interesting that he could have easily afforded a brand-new Roller but chose not to. His comment gave me some insight into a different mindset where people could make choices about how they presented themselves to the outside world.

And me? I didn't know what I wanted. When he asked if I had thought about studying medicine, I replied flippantly, 'Oh no, I don't like blood.' The reality was that I hadn't the slightest idea what to study. I wasn't dismissing his advice outright to make a point — I just didn't really listen to him. It was more that I was drifting in general, and the random currents were pushing me towards a science degree.

So what made me decide to study science?

From my perspective back then, there were three 'types' of science: school science, popular science and science fiction.

School science was sensible and logical — very much basic and essential knowledge that was taught in a problem-solving way. Hardly mind-blowing!

My only source of 'popular science' was *Reader's Digest*, which in its glory days was the best-selling monthly journal in the world. Each edition ran about thirty little stories (one for each day of the month) selected from a wide range of topics, ranging from science to terrible jokes. I can still pull the weirdest pieces of information from my memory banks courtesy of *Reader's Digest*. Once I came up with a crossword answer for 'old bus' (charabanc) because I read it in *Reader's Digest* decades earlier!

But it was science fiction that totally sucked me into studying science at university – I couldn't get enough of it. I especially loved Isaac Asimov, who wrote and edited over five hundred books and is the twenty-fourth most translated author on Earth. His *Foundation* and *Robot* series incorporated psychological storylines, so because of sci-fi, I knew that 'psychology' (whatever it was) had a certain power to influence human behaviour.

So it was mostly thanks to Asimov that I enrolled in a science degree with units in physics, chemistry, maths – and psychology.

At school, if you didn't do your homework every night you got the strap the next day. This meant that I did my homework every night so I didn't get punished, plus it had the added benefit of good marks. But there was no external pressure to keep up with work at university at all, which led to some suboptimal marks for me!

I coasted all the way through this first run at uni life. First-year physics and chemistry were mostly a recap of what we'd already done in high school. But psychology was a totally different box of frogs. I really loved psychology, studied hard and got good marks. In fact, I was so confident about passing Psychology 101 with flying colours that on the day of the final exam, I went to visit a friend in the morning. I wasn't even worried about doing a last-minute swot! I turned up in the afternoon to do my exam completely relaxed, but to my horror I found that the exam had been (and gone) that morning!

There was no way to re-sit the exam, and there was no way to continue further in psychology unless I repeated all of Psychology 101 again. Nope, I didn't love it that much! In retrospect, maybe psychology would have been a really good course for me. 'Reading the room' in an emotional context is not something that I am especially gifted at, so I reckon I would have benefited from the insights that formal study might have provided me.

So instead of psychology, I floated through second- and third-year physics and maths getting pretty terrible marks all the way. I did just enough study to pass but certainly did not acquire the full understanding of physics and maths that I should have. (I've made up for some of my deficiencies since.)

So what was I doing if not studying?

Once I left my single-sex, religious high school, I basically spent a lot of time at the only coffee lounge in downtown Wollongong. I chatted to everyone about absolutely everything – except uni work! All while enjoying exotic foods like iced coffee and toasted cheese sandwiches. I even found a girlfriend! (We hung around as a couple for a while, and then she married my best mate!)

In the university student magazine, *Tertangala*, I was described as a 'left-wing right-wing atheist religious communist capitalist – until he got a blue Volkswagen'. Yep, basically I was confused and unpredictable, but I loved driving! If I wasn't in the coffee shop, I would be driving through the mountains at the back of Wollongong.

I did get very political at university, and it made sense that my first foray into student activism was joining street demonstrations against nuclear weapons. In my first year at uni in 1965, the Cuban Missile Crisis of 1962 was still fresh in people's minds. The United States and Soviet Union were digging in their heels, and the Cold War between them was very rapidly getting colder and more intense. Until 1969, the US had a rotating fleet of nuclear-armed B-52 bombers flying 24/7 just outside the Soviet Union, ready to rain death at a moment's notice. Even as late as 1992, the nuclear-armed US Air Force Bomb Wing squadrons had at least a third of their fleet of planes on high alert, ready to be airborne within a few minutes.

I knew there were some tens of thousands of nuclear weapons on the planet. And *Reader's Digest* (which I still read back then) did its best to ramp up the general paranoia about communism and the evils of the Soviet state. One *Reader's Digest* article proudly described in fine detail the fantastic technology that we in the West had to protect our freedom – boasting that mutually assured destruction (MAD) was the price of that freedom. The doctrine of MAD was, incredibly, the official and public policy of both the USA and the Soviet Union, which meant that a nuclear war would result in the destruction of most of the humans alive on Planet Earth. That really scared the heck out of me!

During my university days, the threat of nuclear annihilation affected me so much that I decided that there was no point in having long-term relationships. I felt an existential emptiness that also precluded having babies. Despite this intellectual stance, I incongruously stayed with the same girlfriend all the way through uni.

But I also got active. I walked in demonstrations, wrote letters to politicians, read incessantly (not about physics or maths!) and talked constantly with my fellow students about all the injustices in the world. I was angry, and I really didn't think it was right that humanity should be 'collateral damage' in an all-out nuclear war.

When I wasn't drinking coffee and protesting against nuclear weapons, I was intermittently protesting to include Indigenous Australians in the census, against apartheid, to free Nelson Mandela and against the Vietnam War. It was a busy time of my life!

Peaceful protest

Protesting peacefully is a distinctly odd activity. Peaceful protest, as championed by Gandhi and Martin Luther King, was very much part of the 1960s zeitgeist and a focus of the news agenda and public discussion.

People protest for many reasons – to publicise their political message, to show concern about a specific issue, to build a bigger base for their cause, to be part of a group, and so on. And sometimes, protests can trigger relatively quick transformation. After George Floyd was murdered by a white police officer in Minneapolis in 2020, the Black Lives Matter protests led to significant changes in policing in the United States.

In September 2023, peaceful climate change protests happened in more than sixty-five countries and involved over 600,000 people.

Today's peaceful protests are very tame compared to those undertaken by suffragettes in Britain in the early twentieth century. Their struggle for women to be allowed to vote (outrageous, huh?) wasn't realised in the United Kingdom until 1928! In 1913, partway along that multi-decade-long struggle, the suffragettes were carrying out some twenty (!!) bombing and arson attacks each month — so back then, it needed more than just 'peaceful' protests to get 'Votes for Women'!

Nowadays, some governments have made non-violent protest a criminal activity. What? There is a global wave of anti-protest legislation combined with increased maximum sentences for activists. It seems to be part of increased governmental powers to declare peaceful protests illegal.

What's wrong with a peaceful protest?

A Periodic Tale

The Vietnam War really opened my eyes. The lottery ballots for who would be conscripted to go to war in Vietnam began just after I started uni in March 1965. I could have been swept up in those drafts, but luckily my birthdate was never drawn.

Communism was the big bogey drawing the United States into Vietnam. The infamous domino theory posed that every nation taken by communism would in turn knock down the nation next door, with an unstoppable tide of communism.

The Vietnam War started as the Second Indochina War (1955–1975). (Indochina then included the countries of Vietnam, Laos and Cambodia. Three million people were killed in Vietnam. But this terrible tally of deaths, inflicted by Western powers, was simply not available until much later.)

The United States had been bankrolling the French in the eight years before the Second Indochina War. They spent some US$3 billion to cover about 80 per cent of France's costs of trying to stop the Vietnamese from getting their independence. The French pulled out in 1954, but the US dug in, spending money like a drunken sailor and increasing the number of US military in Vietnam. By 1969, US personnel serving in Vietnam had skyrocketed to over 500,000. At the end of the war in 1973, over three million Americans had fought, with some 60,000 of those dying there.

In total, about 60,000 Australians served in Vietnam. Right from the beginning, I didn't want to be one of them, but I was also fascinated by the conflict. I wrote letters to the Australian Defence Force and the embassies of the United States and Soviet

Union in Canberra asking for any information on the Vietnam War. The US embassy sent me an envelope a few centimetres thick stuffed with literature. I read every page (and yes, I know, I should have been studying my university lessons!) and also found a revealing Pentagon White Paper analysing the military strength of the communist Vietcong. A subsection looked at weapons captured from the Vietcong. I was surprised to see that most were old weapons – they dated from the Second World War (1939–1945) and some even all the way back to the late 1800s! There was virtually zero evidence to support the common claim that the Vietcong were being armed with new weapons by China or Russia.

Realising that those claims about the weapons didn't seem to stack up led me to wonder about whether the Vietnamese were really trying to spread communism into Australia. Maybe they were simply trying to boot out uninvited military forces to get their own country back! I became deeply suspicious of official Australian and US government material. And I spent a vast amount of energy getting politically educated, and being politically angry, rather than studying my university subjects.

The downside was that I didn't learn more physics at that time, especially Maxwell's equations of classical electromagnetism, quantum mechanics, Einstein's relativities, electronics, and so much more. If I could have mastered these early, then I would have understood so much more. (I would have liked more maths in my brain as well!)

There was one really worthwhile thing that came out of my first bachelor's degree. I did just enough physics and maths to accidentally load a 'mental toolbox' that still acts perfectly as a bulldust filter! Both subjects give you a different way of looking at reality, teach you how to not get fooled and help you see what's really going on in the physical world around you.

Ironically, the main reason I stayed with a straight science degree was simply that I'd missed the Psychology 101 exam! My simple misreading of the timetable shows how little accidents can have big results. Chaos theory (also called the butterfly effect) tells us that small changes in starting conditions can cause massive changes later on.

Regardless, there I was, finishing uni with my degree in hand, just like a little butterfly fluttering off to a new stage of life.

Chaos theory

Chaos theory hit popular consciousness with a 1972 paper by the mathematician and meteorologist Edward Lorenz, entitled 'Predictability: Does the flap of a butterfly's wings in Brazil set off a tornado in Texas?'.

In popular science, this is described as 'the butterfly effect'. Essentially, it says that a tiny change that happens in one location, can set off a major change elsewhere. So if that butterfly in Brazil flapped its wings some three seconds later, it could lead to a storm hitting Louisiana instead of smashing Texas to pieces.

From another point of view, chaos theory looks at what used to be considered total random disorder. Now we realise that these states of disorder can sometimes be predictable. Today chaos theory is used in meteorology, pandemic planning management, environmental science, economics, and much more.

1968–69

The steelworks

At that time in Wollongong, the steelworks was the obvious place to get a job after you'd finished your education. So even though I was busy making myself totally 'individual', I still followed the path of least resistance. In 1968, at the age of nineteen, I found myself employed as a physicist in the central laboratory at Port Kembla Steelworks. I kept trying to enlighten myself about Big Picture Stuff such as politics and the world, but for income, I was stuck in a groove that was well worn and led straight there. Typical for me, I complicated my life at the steelworks by simultaneously studying a non-degree year in astrophysics (just for fun).

Steel is a wondrous material. I love it to pieces, especially stainless steel. Humans have been making steel for about 4000 years, and today there are over 3500 different grades of it. These cover everything from piano wire to the many different stainless steels, tool steels hard enough to drill through stainless, all the hundreds of different grades of structural steels, and many more that you've probably never heard of. Today we make about 1.2 billion tonnes of steel each year. It recycles very easily, and even with our current lazy recycling practices, we still recycle about 60 per cent of what we make.

The steelworks at Port Kembla is just 3 kilometres from the Wollongong CBD, and it's huge, about 8 kilometres from one end to the other. It is so big that while I was working there, a locomotive was lost inside it! One team put it in a holding shed for routine maintenance and forgot about it, then another team tore up the rails that got it there, and it took six months for yet another team to find it.

Besides losing trains, the steelworks had other inefficiencies. At one stage they needed to build a steel bridge from inside the steelworks across the six-lane highway on its perimeter. Strangely, it turned out to be cheaper and quicker to buy the steel from Japan instead of making the steel right there, even though the iron ore and coal to make it had to be transported from Western Australia and Wollongong itself to Japan!

The steelworks took in iron ore and coal and turned them into iron and steel. It had its own large port to unload Western Australian iron ore from oceangoing cargo ships. The coal arrived via a private rail line from the mountains behind Wollongong, as well as from the cargo ships. The steelworks had five separate blast furnaces that turned iron ore into iron. Indeed, Blast Furnace Number 5 briefly held the world record for producing more iron in a 24-hour period than any other blast furnace on Earth.

My job was entry-level and routine, measuring aspects of the steel being produced. It took me a while to get an idea of the big-picture stuff with regard to the steelworks.

Pollution

In my early days at the steelworks, one of my bosses told me, 'The solution to pollution is dilution.' In other words, don't bother to clean up your process or waste products – nope, just dump it all in the Pacific Ocean (which, by itself, is bigger than all the land masses on Earth combined). Actually, the cheapest option was to dump waste straight into the harbour at Port Kembla and let it drift out to sea with the currents and tide. This was pretty shocking to me – even fresh out of uni I knew that pollution was generally bad!

They didn't mind dumping in their own backyard either. There were waterways inside the steelworks because it was built around a natural port. One of my colleagues once had the job of taking an aluminium boat up these various waterways. He told me the water he tested varied in colour from bright green to bright orange, and from very acid to very alkaline. Regardless of the colour, the water would eat holes in his aluminium boat within a year. That was not a natural action of regular unpolluted water!

How to make steel with hydrogen instead of carbon

First, let me start at the very beginning – with atoms.

Every physical thing is made from atoms. In chemistry, iron atoms have the chemical symbol Fe, oxygen atoms have the symbol O, while carbon is designated as C. So iron ore, which is a compound of iron and oxygen atoms, is shown as FeO.

Second, let me give you a summary of three rather different substances – iron ore, iron and steel.

Iron ore comes from the ground. It was 'manufactured' a few billion years ago, when photosynthesis began to take off. The oxygen atoms produced by photosynthesis combined with raw iron atoms to give iron ore (or rust, if you want to be brutal). There was so much oxygen around in the atmosphere that all the iron atoms ended up stuck to oxygen atoms.

Iron can be made from iron ore. If you strip off that oxygen atom from the iron atom, you end up with pure iron (freed at last, after a few billion years). Iron is just a collection of iron atoms. It's useful, but it has disadvantages – it rusts, it's not really that strong, it's hard to make 'springy', it's a bit difficult to weld, it's hard to drill into, etc.

Steel is iron with small amounts of added carbon – somewhere between 0.02 and 2.14 per cent by weight. Those carbon atoms remain forever locked up between the iron atoms. They do *not* contribute to climate change/global warming. (You can also add other elements to further improve the steel – vanadium, chromium, nickel, etc.) This relatively small number of carbon atoms very dramatically changes the properties of the final product (steel). Steel is much more useful than iron (and in fact, absolutely essential to our civilisation). This incredible versatility drives our civilisation to make steel, instead of sticking with the less useful iron.

Third, here's how we usually turn iron ore first into pure iron, and then later, into steel.

We start with iron ore – which is FeO.

If you can get rid of that pesky oxygen atom you'll end up with relatively pure iron. The principle is fairly simple. Start by grinding coal into a fine powder (to give you carbon), and then grind iron ore into a fine powder. Blow lots of very hot air through that mixture. Once the mixture gets to around 900°C, a chemical reaction kicks off. The carbon grabs the oxygen atom off the iron, forming carbon dioxide – and leaving behind relatively pure liquid iron.

$$2\ FeO + C \rightarrow 2\ Fe + CO_2 + Heat$$

There are two bonuses in this process for the steelmaker.

First, the carbon dioxide vanishes into the atmosphere and becomes somebody else's problem (at the moment, our shared air is a free garbage tip for some polluters). Unfortunately, it's bad for the rest of humanity – steelmaking accounts for about 8 per cent of the world's carbon dioxide emissions. This is obviously a significant component of climate change/global warming.

Second, the reaction gives off huge amounts of heat. Yes, you do have to provide some heat at the beginning to get the reaction started, but once it gets going, you don't have to add any more. All you need is to keep adding more finely powdered iron ore and coal, and every now and then, open a valve at the bottom, which will let the liquid iron run out. So in this case, the heat produced is an added bonus.

Then later, while the liquid iron is

still hot, you can add extra carbon to give you the precise variety of steel you want.

Now here's the big surprise. We already are making steel without producing any carbon dioxide, or any other greenhouse gas emissions at all.

Making steel doesn't have to involve burning carbon! You can burn hydrogen instead. You can get your hydrogen from water, using renewable electricity.

$$2H_2O + \text{Electricity} \rightarrow 2H_2 + O_2$$

In traditional steelmaking, carbon grabs the unwanted oxygen atom from the iron ore, and gives off carbon dioxide and heat. Hydrogen can do the exact same job, but without the CO_2.

$$FeO + H_2 \rightarrow Fe + H_2O + \text{Heat}$$

In Sweden, the steelmaker SSAB has already teamed with Volvo to make automatous electric mining dump trucks with steel made using hydrogen. Sweden tends to be a world leader in many ethical manufacturing processes. They realise that climate change has an economic cost (currently 1 to 2 per cent of the world's gross domestic product)

> and are willing to invest now to make
> profits later.
>
> I dream that my old steelworks at Port
> Kembla will soon embrace new technologies!

Non-physics stuff at the steelworks

About six months after I'd started working at the steelworks, I had a sudden overpowering urge for a meat pie. I hadn't brought any lunch in, so I walked about half a kilometre to the canteen. And while I ate my meat pie, the most momentous thing happened: on the black-and-white TV in the small canteen I saw the first humans walking on the surface of the moon. It was 21 July 1969, at 12.56 pm. When anyone asks where I was when the first moonwalk took place, I have to admit that I was in the steelworks canteen eating a meat pie. I still get a flashback to the smell of pastry and tomato sauce when I think of that first moonwalk!

One fun thing about the steelworks was joining the Steel Industries Auto Club. But looking back, the monthly Friday night car rallies were a little dangerous (perhaps more than just 'a little').

Each event was a midnight-to-dawn rally, with several high-speed sections (usually on dirt) separated by transit sections (usually on bitumen). I didn't think about it at the time but these rallies were probably illegal. I don't recall the organisers getting official permission to drive at supersonic speeds on the deserted back roads, but perhaps they did. Certainly I never saw any roads closed off from the public for the high-speed sections. Anyways, as it turned out, nobody else was driving on those quiet back roads between midnight and dawn.

Each car had a driver and a navigator. The route was different for each event, and wasn't marked. Each time, we would be given a printed set of slightly cryptic instructions that the navigator had to interpret (e.g. '4.2 miles after the river crossing, turn right at the first opening in the barbed wire fence').

Most of the other driving teams had hotted-up cars, like factory-modified Mini Coopers and Volvo rally cars, and Falcons and Holdens with the biggest V8 engines they could buy and then modify. All I had was a crappy blue Volkswagen with a gutless 36-horsepower engine that couldn't pull your granny off a dunny. I wasn't sparklingly good as a rally driver, but then neither was my car. But I was excellent as a navigator, and happy to team up with someone else. It was so much fun.

We'd finish on a beach at dawn with a picnic. One trick I quickly learned was to buy a pre-cooked chicken, wrap it in aluminium foil and tie it to the exhaust pipe with steel wire to keep it warm. On the beach, after several hours on the exhaust pipe, it was still hot and delicious. I never got food poisoning, but I don't know why I wasn't worried about what the exhaust fumes did to my chicken breakfast!

> ### Torque is tight and right
>
> I remember being told a story with a life lesson by one of my bosses.
>
> The steelworks bought a huge machine from Japan called a rolling mill. It measured

about half a kilometre in length, and could turn a 10-tonne rectangular block of red-hot steel into a sheet some 3 metres wide and 100 metres long! The Australian engineers installed it, but it just didn't work consistently or reliably.

The steelworks executives complained to the Japanese, implying that they'd sold them a bad mill. They must have touched a nerve of national pride because the top executives flew in from Japan almost immediately.

These top bosses and engineers put on pristine white overalls and, using giant spanners and torque wrenches plus cranes and jacks, they very slowly and deliberately started at one end of the mill and completely dismantled it. Every single individual component was separated from its neighbouring component. The whole rolling mill was then put back together again, and along the way, tightened up to the exact specifications in the manual. They didn't install any new parts – they simply reinstalled the whole mill, but this time according to the exact specifications.

When they finished, they tested it. The steel went through and the rolling mill did exactly what it was supposed to do. The

> Japanese executives then shook hands with everybody and went back to Japan.
>
> The THM (Take Home Message) was that if you have a job to do, do it properly.
>
> I told my father this story and he immediately responded with a Latin proverb: *Idem est facere non est idem.* It translates to 'Two people do the same thing, but it is not the same thing.'

Physics stuff at the steelworks

I got my job at the steelworks without having any background in metallurgy (the science of metal). But as a physicist (even with lousy marks at uni), I still had the mental toolbox to extrapolate from basic principles, and then load into my brain any specific knowledge relevant to a new field. The problem-solving aspects of your mental toolbox help in virtually any situation. They let you jump into any job – geology, biology or, in my case at the steelworks, metallurgy. All I had to do was to load my brain up with the data related to that field, and apply my mental tools.

So this job was my first introduction to the world of metals.

At the start, I had no idea that you could make one metal harder than another, so that the hard metal could cut the soft metal. For the first time, I realised this was the principle behind the hacksaw, the chisel and the spinning metal drill that can cut through other metals. I learned a lot on the job, studied up and asked for help from experienced staff whenever I needed it.

One of the properties of steel is its flexibility. If steel is made correctly, you should be able to bend it at least a bit. One of the metallurgists called me into their lab to show me something amazing. They said, 'Look at this,' and dropped a piece of steel about the size of an A3 sheet of paper and 2 centimetres thick from a height of a couple of metres onto the ultra-hard steel floor of a protective cage. The steel shattered like glass! It was at the extremely 'brittle' end of the flexibility scale. Imagine if that was in your staircase! That's when I realised that sometimes bad steel got made accidentally and, even worse, sold.

My job was not fundamental research; I was much too junior and didn't have enough knowledge. Instead, I was in a 'service' position. My specific job was to measure the fatigue limit of the various steels that were being produced. 'Fatigue' is a concept that is very important in engineering, for example, to ensure that bridges stay standing 'forever'. The fatigue limit becomes critical after a long period of use, not in the short term (unless the engineers have got their calculations very, very wrong). If you buy a car with lots of kilometres on it, occasionally some moving part that looks perfectly fine will break with no warning. That part has reached its fatigue life – and you have to repair it.

Speaking of bridges, one of the very large jobs the steelworks took on while I was working there was supplying steel for the West Gate Bridge in Melbourne, which now crosses the Yarra and heads west out of the city. The West Gate Bridge infamously collapsed during construction, and that's definitely not a good thing.

A Periodic Tale

The West Gate Bridge collapse

The remarkable thing about the West Gate Bridge – apart from it collapsing during construction – was that it was one of the largest box girder bridges in the world at the time. Box girder bridges were invented as recently as 1919, specifically to make light and strong temporary bridges that could be used on the battlefield so heavy tanks or trucks could cross a waterway (or any other gap).

> **Box girders**
>
> A 'girder' is a major horizontal beam that can support other smaller minor beams. A 'box girder' is a horizontal beam made from a thin wall of strong steel shaped into a long box, with the ability to carry the specified load. This design gives lots of strength without being too heavy.
>
> Imagine a single sheet of A4 paper. One single sheet of paper is weak and floppy. It can't support any significant weight at all. But what if you fold the paper in half lengthwise, and then fold it in half lengthwise again? Unfold, then join the two long edges together with sticky tape. Suddenly, you've got a skinny box made of very thin paper that is surprisingly strong.

Dr Karl Kruszelnicki

> This box design is the basis of the box girder construction used for the West Gate Bridge. Instead of paper, a thin wall of strong steel is shaped into a box with the ability to carry the specified load.

The findings of the Royal Commission into the collapse of the bridge in 1971 laid the blame on both the original structural design and the construction method used by the original builders. This bridge build was pushing the limits of design technology at the time. But here's an anecdotal list of other problems as told to me by colleagues in the industry.

First, computers at that point couldn't simulate the performance of a bridge *before* it was built. Nowadays computer modelling would make this an easy task. But in 1969, the West Gate Bridge was around the limit of box-girder-bridge engineering. Back then, engineers followed the age-old tradition of making the new structure (cathedral, bridge, tower and so on) just a little bit bigger than the last time. At some stage the structure will fall over, and that's clearly the marker that you've gone too far!

Second (and remember that this is hearsay), the steel for the bridge was made in two grades, stronger (LY50) and weaker (LY35, about a third weaker in strength). Apparently, in the early days of the project, the different grades of steel were not properly labelled when loaded onto the train at Port Kembla. (The labelling was updated for later steel shipments.) I was told that early on at the job site in Melbourne, the construction team just grabbed the nearest lump of steel and chucked it in wherever. Using high-strength steel in a low-strength application was no problem – just a bit of overkill. But putting the low-strength steel into a high-strength application had the potential to lead to Bad Things.

The third complication was the serious one that led to the collapse. The box girders were constructed on the ground in sections and then lifted into position by cranes to be welded in place. The section that broke was between piers 11 and 12, which were about 50 metres high. Unfortunately, the box girder ends (one coming from pier 11, the other from pier 12) didn't meet at the same level: instead of 'kissing', they were about 114 millimetres out of vertical alignment with each other.

Misalignments are common in construction, and there are usually a few ways to adjust things, with some choices better than others. The decisions made by the construction team in this case led to disaster. At 11.50 am on 15 October 1971, the box girder between piers 10 and 11 broke and the whole

2000-tonne section smashed to the ground. The impact was so loud that it was heard 20 kilometres away. Tragically, thirty-five workers died.

All work on the West Gate Bridge stopped immediately. It was an industrial tragedy. In fact, around the world, three other box girder bridges collapsed during construction between 1969 and 1971, which led to a major reappraisal of how to build them.

It took a Royal Commission and ten more years to finally complete construction of the bridge.

Testing the steel for the West Gate Bridge

The collapse of the West Gate Bridge terrified me. I had already resigned from the steelworks when the bridge crashed down in 1971, and was far away, working in New Guinea. But a lot of my time at the steelworks had been spent doing fatigue testing of the West Gate Bridge steel. It was an incredibly minor role, and totally unrelated to the bridge collapse, thank goodness, but I didn't know that at the time. There was a moment of dread when I heard about the bridge collapse that I could have been somehow to blame. You see, testing that steel had led to a confrontation with my boss that ended up with me resigning.

Fatigue limit testing is essential if the metal is going to be in place for a long time and go through repeated cycles of stress many thousands or millions of times. Previously, it was not standard practice at the Port Kembla Steelworks to test fatigue properties of steels. I was brought in specifically to set up the fatigue limit testing unit.

In construction, the concept of 'fatigue limit' relates to whether the metal will eventually break after stresses are repeatedly applied to it. The stress could come from heavy trains rolling across a steel bridge, or from water being churned by the steel propellor on a big ship, or from the changing air pressure on the fuselage of a jet plane as it flies from ground level to 41,000-feet altitude and back again. The metal is matched to the task – and is designed to endure and be able to carry impressive loads indefinitely. A fundamental part of that infinite life span is that all stresses need to be kept below the fatigue limit. If you can do that, the fatigue life is infinite.

Fatigue limit

```
Do this fun little experiment.
   Get a paper clip, straighten it out, bend
it back on itself a full 180 degrees, and
then make it straight again. You can do this
once and it won't break, but if you bend and
straighten it repeatedly, around the fourth
or fifth time it suddenly feels strangely
easy to bend or straighten. Around the
seventh or eighth time it will unexpectedly
break into two pieces. On each bending
cycle, you've stressed the paper clip metal
above its fatigue limit, so the fatigue life
is only about six or seven cycles.
   Now straighten out another paper clip
and do the whole bend and straighten thing
```

again, but this time bend it less, say, to 90 degrees. This time it might last fifty bends before it breaks. At the lower stress, the fatigue life is fifty cycles. In the graph, as the stress (on the vertical axis) reduces, the number of cycles (on the horizontal axis) increases.

Then do it again but reduce the bending angle further (say, to 45 degrees) and it will last perhaps 500 bends. Keep going – the lower the bending (or stress) that you put on the steel, the longer it will last. The number of cycles increases to 5000, then 50,000, and so on. Interesting!

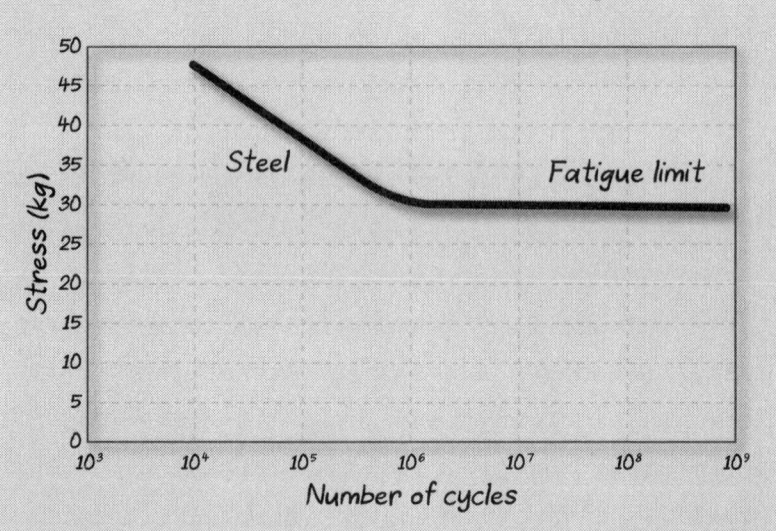

Graph: Stress (kg) vs Number of cycles (10^3 to 10^9), showing Steel curve declining to a Fatigue limit around 30 kg.

(Source: Andrew Dressel/Wikimedia Commons)

The weird thing is that when you lower the repeated stress enough, the steel will never break. That's right – never. So if

A Periodic Tale

> the stress is below the fatigue limit, the steel will in theory last forever. Note how the steel curve on the graph suddenly goes horizontal at around 30-50 per cent of the maximum strength of the metal.
>
> The fatigue limit depends on a combination of both the stress you apply and the properties of the metal. If you build a steel bridge, you want the repeated stresses in every steel component of the bridge to always be below the fatigue limit so all steel components will never fail. And the bridge will stay standing!

So here I was in my new job as the fatigue life officer. (The title didn't relate to me always being tired from having too much fun — and anyway, from a philosophical point of view, there's no such thing as too much fun!) The position was so brand new that I was going to have to build the testing machines from scratch.

First I had to learn what the heck fatigue limit was (I had no idea when I started!). Then I learned why it was different in different materials (for example, steel versus aluminium) and how scientists measured it. Then I got help (lots of it!) to design and build different types of testing machines. I asked for steels with known fatigue limits and then tested those in my machines. Hooray! My machines got very similar results, which meant I was able to confirm their accuracy. I was now the 'King of Fatigue' (at least in Wollongong) and ready to rock.

In 1969 and 1970, I was measuring the fatigue limits of the two major steels that were going into the West Gate Bridge. It was a big job. Several months later, my data was showing a reproducible and very concerning pattern: neither of the two steels met their specified fatigue limit! From memory, they were too weak by a small margin.

I took the results to a supervisor and told him my concerns. He didn't even look at the paperwork showing the results. He just looked me in the eye and said, very deliberately, 'Karl, I wouldn't be upset if you went back and checked your numbers and found that you'd made a mistake.'

I was very green and took him at face value. Good idea, I thought – measure twice, cut once! It's always worth double-checking your numbers.

I spent the next week rechecking every single part of the testing procedure. I found a few minor mistakes, but nothing significantly changed the data about the steel's fatigue limit being just under what it should have been.

I came back to the supervisor and said, 'Look, it's still too weak.' Everything played out again in exactly the same way: 'Karl, I wouldn't be upset if you went back and checked your numbers and found that you'd made a mistake.'

I was mostly confused by his response. I suggested some fellow physicists confirm my calculations and got authorisation to pull them off their own work to do so. They found the same result – there were no mistakes.

I went back to the supervisor for a third time, but it was like a broken record. Suddenly, my naïve 21-year-old self realised what was going on: he wanted me to fake the results!

This steel was being made in absolutely huge quantities by the steelworks, and I suspected he wanted to avoid the mess of having to recall it all. To be fair, perhaps he wasn't convinced about my testing because it was new to the steelworks. It would definitely have made his life easier if he could just say the steel met all the specifications required.

> **Could the West Gate Bridge still collapse?**
>
> There is the remote possibility that the West Gate Bridge will collapse in the future due to various fatigue-related issues. But it's highly unlikely. It would require several contributing factors, in a specific sequence. In aeroplane disaster investigations, the ultimate crash is hardly ever caused by a single factor. Instead, the crash is usually caused by a sequential series of small mistakes.
>
> One such mistake would be if the fatigue limit of both steels (LY35 and LY50) is too low. But if this were the case, it would only be slightly below specification. There should be enough of a safety factor built into the engineering process to allow for this, making this slightly inadequate specification not really relevant.

> Another possible mistake would be using the low-strength steel in a high-strength application. Suddenly, the steel is about one third too weak. But again, the safety margins would likely protect against a failure.
>
> So while anything is possible, it seems very unlikely that one day the West Gate Bridge will plummet to the ground.

I was at an impasse. Then three things happened in rapid succession.

First, I wouldn't change the results, which meant life at work became quite awkward and unpleasant.

Second, my girlfriend left me to marry my best mate. For me, the break-up came out of the blue, and I was devastated. I almost certainly wasn't a very good boyfriend if I didn't know she was so unhappy and wanted to marry my best friend! As for my best friend, I guess he needed to be doing what made him happy, not me. I didn't keep in touch with either of them after the break-up.

Third, I resigned. On a whim, I decided to head off to New Guinea, where unbeknown to me I had a similarly difficult boss awaiting me!

Part 2

The drug-crazed hippie years

The word 'career' comes from the Latin *carrus*, meaning 'chariot'. It can be defined as 'a person's pathway, or course, or progress, through all or part of their life'. This sounds stately, planned, organised and controlled. But there's another definition that means the exact opposite of totally controlled, as in 'he careered wildly out of control down the steep hill after both the brakes and the steering failed'. My career and I embody both definitions of that word!

In my early twenties I was entering my drug-crazed hippie era. This period of my life saw enormous changes. Instead of just one scientific career, I now had four very different side-by-side careers running in parallel. I was splitting my days between working as a filmmaker, being a roadie for rock 'n' roll bands, driving a taxi around Sydney, and picking up mechanical, electrical and electronic skills.

The brutal truth is that I wasn't a 'success' in any of these endeavours (except maybe taxi driving). But everything I learned and loaded into my mental toolbox (that phrase again!) in this time turned out to be 100 per cent essential to my later careers.

1970–71

A hippie in New Guinea

The times sure were still a-changin', and I was changing with them, or at least I was trying to.

In August 1969, the Woodstock music festival exploded into the collective consciousness. It was the biggest music festival in history, with about 460,000 people attending to see thirty-two acts, which climaxed with Jimi Hendrix slaying the audience. Woodstock was one of the defining events of the twentieth century that rocked the counterculture into the mainstream. I didn't even know what counterculture was, but I loved the music, and I wanted to be part of it.

Dropping out – from the steelworks

While still at Port Kembla, I read an article in the December 1969 issue of *Scientific American* called, very simply, 'Marihuana'. It stated, 'There is considerable evidence that the drug is a comparatively mild intoxicant. Its current notoriety raises interesting questions about the motivation of those who use it and those who seek to punish them.'

I was very interested, but I had no idea where to find some of this 'marihuana'. So over the next six months I did the next best thing and went about trying to at least *look* like a hippie. I bought some bell-bottom jeans and stuck giant adhesive flowers to the doors of my VW Beetle. (That got a bit of gentle ribbing at the steelworks.)

I also wanted out of the steelworks, and away from my boss. I saw a newspaper ad for a tutor at what the locals called the Lae Institute of Higher Technical Education, a tertiary college in New Guinea. I knew nothing about New Guinea, but I wanted to get as far away from my life in Wollongong as possible – I was keen to move on from my ex-girlfriend, ex-best friend and soon to be ex-job! New Guinea also sounded pretty exotic compared to Wollongong, and it was relatively easy to get to and live in because in those days it was still governed as part of Australia.

I got dressed up to look 'professional' for the interview – clean-shaven with short hair and glasses, wearing a neat light-grey suit and white shirt with a tie – and drove to the University of New South Wales to meet the head of the department at Lae. I got the job and a few months later headed off to New Guinea.

But the 'new me' who turned up for work in 1970 was a bearded bloke with intense eyes (resulting from contact lenses combined with extreme short-sightedness), a leather headband holding back my unkempt long hair and wearing a then-popular string vest. (I later found out that this led to my nickname of 'Nudge Nipples'.) I turned up to the HiTech, as the campus was also known, in New Guinea and my new boss was floored, saying, 'You look nothing like the guy I picked at the interview.' He was 100 per cent correct!

New Guinea 101

Humans first settled Papua New Guinea at least 42,000 years ago. It was physically separated from the Australian mainland by the rising ocean about 10,000 years ago as the Earth cycled out of its most recent Ice Age. Various European powers have claimed control of New Guinea, but it came under Australian administration during the First World War and then gained independence in 1975.

Today, New Guinea is one of the most linguistically diverse nations on Earth (there are about 839 different languages) and the least urbanised (only about 13 per cent of the population live in cities).

When I lived in New Guinea, kiaps, or district patrol officers, were in effect the law. (The word comes from *Kapitan*, German for 'captain' – the Germans had their turn at controlling part of New Guinea prior to the First World War). Between the two wars,

> kiaps did training in tropical medicine, anthropology and law at the University of Sydney before they went to New Guinea. They went into remote areas and were virtually a little mobile government in their own right. Part of their role was to introduce a Western style of law and order, with the goal of shifting the culture away from traditional payback killings.
>
> Kiaps had enormous power, and the more isolated the region, the greater their power. On occasions they would have the combined powers of a police officer, lawyer, judge and jailer. They could even be a medical officer and dentist as well as agricultural adviser, surveyor and engineer.
>
> Kiaps were once the stand-alone representatives of the Australian government, but their roles and powers changed as New Guinea moved closer to independence.

In 1970, when I arrived, the Australian government was still running Papua New Guinea. Independence was half a decade away.

The HiTech, which was my new employer, was in transition. By 1973 it was renamed the Papua New Guinea University of Technology, but I was long gone by then.

The Australian expatriate staff I met in New Guinea

mostly wanted to help locals get a better education and better standards of living, but a small number were third-rate workers who would never have made it to any position of authority in Australia. I tried to avoid mingling with this subset of embittered expats. Even though I had left Australia feeling pretty low and with my tail between my legs, I never felt bitter.

Working in New Guinea, I had both a relatively high income and a lower tax rate than in Australia. Imported goods were even cheaper because they were virtually (if not completely) free of tax. I felt like I had money to burn, so having always been interested in audio and photography, I bought my first proper single-lens reflex camera, a high-quality audio system, and a 16-millimetre Bolex movie camera. I bought a Land Rover, so I could leave Lae on the weekends to visit different villages. The locals were always very kind, and I photographed and filmed everything.

I joined the film club at the HiTech and saw all their European 'art' movies, truly falling in love with film for the first time. At the local Lae movie theatre, the wealthy people sat downstairs for ninety cents, or upstairs for a dollar. I sat with the locals in a deck chair right at the front for ten cents. That really let me soak up the visual experience, by being as close to the screen as possible. My terrible short-sightedness is the other reason I still prefer the front row!

The food market in Lae was eye-opening. The surrounding Morobe Province is famous for its taro, bananas, sugar cane and so many other fruits and vegetables. I had never seen such an amazing array of fresh and tropical fruits anywhere back in Australia.

The students at the HiTech came from all over Papua New Guinea, with varying standards of Western 'hygiene'. So all the enrolling students got sprayed for hair lice. We cut ping pong balls in half, put one over each eyeball, and sprayed their thick hair with DDT! I shudder at the memory of this practice – but I didn't know any better at the time. DDT was developed as an insecticide and had terrible environmental impacts. Rachel Carson's 1962 book *Silent Spring* drew attention to the impact of DDT, especially on birds. It is definitely *not* safe for humans and is now mostly banned around the world.

Anyhow, I loved my job as a physics tutor. It was tricky teaching basic physics concepts, such as acceleration, to people who had never been in a lift or plane and who walked everywhere. They sometimes had been in a car only a few times in their whole lives. This was just one example of how the staff had a totally different set of background reference points to the students. The boss was always 'busy' so I started giving his physics lectures as well as running my tutorials.

I wasn't trained as a teacher, so my style was a bit unorthodox. For example, I came up with a very left-field way to teach my students about different types of friction.

Luckily, nobody got injured.

Wild wheelies in the parking lot

Friction is the force that resists movement between different surfaces. There are two main types of friction, and I wanted to teach my students about this.

First, there's static friction, which is what you have to overcome when you try to get something to shift from being stationary to moving. For example, think how hard it is to slide a fallen log off a concrete footpath.

Second, there's dynamic friction (or sliding friction), which is the force needed to *keep* our log sliding across the concrete. Once your log is already sliding, you can move it more easily.

The basic point I was trying to get across to my students was that dynamic friction is always less than static friction. Usually, this is shown or demonstrated on a laboratory bench, with some blocks of wood, pulleys, strings and weights.

Not me.

I wanted to do handbrake turns in the HiTech parking lot.

I had one student at a time sit in the front of my Land Rover with me, both of us wearing a seat belt (safety first!). Then I did wheelies (circles, doughnuts, whatever you want to call them) in an empty section of the parking lot at a high speed – just below the speed where it would tip over. At this point, there was static friction between the wheel and parking lot surface, because the wheels were rolling over and gripping the surface. You could tell it was static friction because there was no sliding. When all the tyres are gripping the ground, you travel in a nice circle, if you keep the steering steady.

Once we were circling at a constant speed, the student would pull hard on the handbrake that sat between us, as I took my foot off the accelerator and pressed on the clutch. The back wheels were no longer connected to the engine and they stopped turning. They were no longer rolling over the road – they were sliding. We were in the Land of Dynamic

Friction – which, remember, is always less than static friction. Less friction on the back wheels meant they didn't stick to the ground as well – so the back wheels of the Land Rover lost their grip and swung out, and suddenly we'd be pointing the other way.

Bingo, your classic Hollywood handbrake turn – with a free physics lesson thrown in!

The situation was 'pretty' safe because I had a lot of experience, as well as a few advanced driving skills. (All that time rally driving in the mountains behind Wollongong wasn't wasted!)

Everyone got two turns in the front seat, pulling hard on the handbrake. The first time was to try it out and be shocked that it actually worked. The second time was to confirm that they could reliably reproduce the phenomenon of the handbrake turn. This reproducibility proved it was Real Physics. The kids found it really exciting, but more importantly they also now had a physical, intellectual and very memorable understanding of these two types of friction.

By the way, it's not a recommended teaching practice. And definitely don't do handbrake turns in a high-mounted vehicle like a four-wheel drive or an SUV. But the parking lot had been recently covered with very smooth bitumen, and I had checked it out the night before, I promise.

Fly me to the moon

At this time, the United States was still flying humans to the moon. (Moon landing deniers? Sometimes a belief is so far out that it's not even close enough to reality to be 'wrong'.) Fairly soon, Newton's three laws of motion came up on the physics syllabus for me to teach, and as it turned out, NASA used all three to get astronauts to and from the moon. So I took the students through the physics of how we humans got to the moon. It was beautiful.

> **Newton's Three Laws of Motion**
>
> First, a body will keep on doing whatever it's doing – if it's stopped, it will stay stopped, and if it's moving, it will keep on moving – unless a force acts upon it.
>
> Second, the force on a body is equal to its mass multiplied by its acceleration.
>
> Third, for every action, there is an equal and opposite reaction.
>
> These laws are easy to say and remember, but carry surprisingly deep concepts within them.

The astronauts launched and got into low Earth orbit (LEO) about 400 kilometres above the ground. Then they accelerated to leave LEO to head for the moon, slowed down to insert themselves into the moon's orbit, and slowed down again to leave the moon's orbit to head for the lunar surface before slowing down even more to finally reach zero velocity at the exact moment their landing legs touched the ground. Later they took off from the surface, got into orbit around the moon, left moon orbit to get to Earth, and finally slowed down to enter the Earth's atmosphere at exactly the right angle and speed. We even did the physics of parachutes (used for the last descent) and the physics of splashing down in the water of the Pacific Ocean.

Three things really stuck in my mind from that experience.

First, after a few weeks of learning the physics of getting the astronauts to the moon and back, the students understood Newton's three laws of motion. That was pretty good!

Second, in teaching Newton's laws for the very first time, I too gained a deeper understanding of them. I had no idea that my previous 'knowledge' had been so shallow.

Third, that when I was teaching Newton's Law of Motion it happened to be an especially humid time of year in Lae (just 6 degrees from the equator). On some days, tiny droplets of water would appear on the blackboard in a line close to and parallel to what I was writing on the board! The microscopic particles of chalk were acting as 'nucleation centres' that made water molecules in the air stick together and land on the surface. I'd never experienced that kind of humidity before!

Things were going really well for me working as a tutor, but cracks were beginning to appear in the complicated interpersonal realm between my boss and me.

Problems with the boss (again)

The first problem at work was that, for no good reason, my boss thought I was having sex with his wife. Nope.

His wife was not an academic, but she dropped into the HiTech a few times each week so she could meet people there and have a cup of tea. I usually had morning and afternoon tea in the workshop with the electronics and mechanical technicians. Sometimes she would drop by and we'd have a laugh and a joke. (But, let me emphasise, no sex.) My boss thought that the laughing and joking were a sure sign that his wife was having an affair with me. Strike one against me.

The second problem was that towards the end of the first year, the science students voted on how good their teachers were. I came first, and my boss came last. Strike two against me.

The third problem related to my boss's PhD research into hair and wool fibres. He had a proposal that hair could be a long-term marker of a person's diet, relating it to the fact that students at the HiTech were no longer living in their traditional homes or eating their traditional diets because they were boarding in college. They might, for example, fall in love with bacon, and no longer eat fruits or vegetables. So he wanted to measure the impact of dietary changes on their hair, and he was planning to do this by measuring the diameter of their hair as a marker of nutritional status.

Sounded hairy to me.

Unfortunately, he wanted me (not him) to do all (not just part) of the work for his PhD. But even worse, he knew what 'results' he wanted to get. So my job was to go through the motions, spend the next two years looking through the microscope at hair fibres and write down what I measured – but on the side there would be a separate set of notebooks prefilled with his fake numbers that would fit in with the results he wanted to get.

I left the steelworks precisely because I thought faking data was unethical and unscientific, and I was most reluctant to play any part. Instead of addressing this issue head on, I just put my head down, and kept on taking and recording the real measurements. So there was extra tension building up between us. Strike three.

Exploring New Guinea

My life inside the university, when I was not with the wonderful students, was becoming maddeningly frustrating. However, I was in the exotic tropics, so I focused more on my outside life and seeing the amazing sights of New Guinea.

I worked hard at my job Monday to Friday. But come the weekend, I got out and explored. I had my first ever four-wheel drive vehicle, which I could use to adventure out into the most amazing landscapes that were virtually on my doorstep.

As an example of how different New Guinea is from Australia, just consider the rainfall. In Lae it rained every afternoon at around half past four, regular as clockwork. I would sit on the verandah with a cup of tea and marvel at the regularity and intensity of the downpour. And the rainfall was truly torrential! The first time I drove out of town I was puzzled to see a steel bridge, 50 metres long and 10 metres above the ground, sitting in the middle of completely dry land, with the nearest section of river several hundred metres away. Turned out that in one incredible downpour, the river had changed course due to the massive waterflow and simply left the bridge sitting high and dry!

One trip I made with one of the HiTech staff was to the Caves of the Dead on the way south to Wau. The local custom was to leave the bodies of the deceased in a cave to dry out, then families would visit the dead relatives, clean the bones and keep the dead up to date with gossip. As I walked along the cliff there were little alcoves full of bodies in various stages of mummification, something I'd never seen before. (Years later we saw similar burial traditions in Madagascar.) There was so much for me to learn about the cultural life of New Guinea.

Running lengthwise along the country like a spine are the New Guinea Highlands, which reach heights of 4500 metres. After the Caves of the Dead, we ended up at Wau, itself over a kilometre above sea level. Today Wau is a small village south of Lae, but back in the 1930s it was one of the busiest airports in the world due to gold mining. Big mining companies flew in vast mining dredges, piece by piece, and assembled them in Wau. When the bulk of the gold ran out, the machinery and equipment were left behind to rust.

In Wau, I met someone who still made a living from gold. His ridiculously simple process was to strategically place long-fleeced sheepskins along a stream that ran through his property. The fleeces collected gold particles washing down the river, and when there was enough, he took the fleece out of the river and burned it. The gold was left behind in the ashes.

How to lose your toenails

For some forgotten reason (maybe a charity event?) I found myself part of a HiTech group attempting to climb the highest peak mountain in New Guinea, Mount Bangeta.

As a certified hippie, my mission was to never wear shoes. On the other hand, there were terrible tropical infections in New Guinea that could enter my body through my feet. So as a compromise, I wore thongs. Now even I knew that thongs were no good for mountain climbing, so I bought some rubberised walking boots. They fitted fine in the shop, so I put them aside until the day we flew out.

Eight of us at a time, including the pilot, flew up a very narrow gorge, with mountains towering above us on either side. It was really quite terrifying! Just as it looked like we were running out of gorge and about to run into a mountain, we turned hard right. Amazingly (although presumably not to the pilot) there was a very short and narrow airstrip right in front of us that sloped steeply uphill. The plane landed and we all piled out with our backpacks, full of enthusiasm. The light plane turned around, headed downhill and literally dropped off the end of the airstrip. It then popped up again, waggled its wings at us to say goodbye and vanished down the narrow gorge.

There were about twenty of us from the HiTech, staff and students who were woefully underprepared in all sorts of different ways. The walk was very hard going – steeply uphill followed by steeply downhill, and repeated over and over. Struggling on, we arrived at a small village around midday where the locals very kindly fed us. By the afternoon we were paying the local women a silver shilling (ten cents) each to carry our bags along with their loaded bilums, the traditional string bag with a broad handle that goes across the forehead so the load rests on the back. Some of the women had been carrying heavy bilums for so many years that they had a permanent indentation in their forehead as thick as my little finger! Incredibly, some women also carried a baby in another bilum on their back; regardless, they took our shillings and then promptly and effortlessly galloped out of sight. Even carrying our extra loads, they still got to the next village a few hours before we did!

We hired locals to be our guides and to translate as we crossed from one tribal zone into the next, which could simply be the valley between two chains of mountains. We could find three different languages in three adjacent valleys and have to swap guides three times!

Very quickly, my toenails went bad. Going uphill was fine, but going downhill, my too-long toenails rammed into the front of my new boots. The pain was terrible. Nobody had any nailcutters, and the machetes the locals had were way too big and blunt. I was getting desperate enough to think about asking someone to bite my toenails short! I couldn't reach them myself or I would have happily bitten them down.

The first two nights we slept in villages, then on the third night we slept just below the tree line on the steep slopes of Mount Bangeta. It rained torrentially during the night, and two of the tents got washed straight down the hill! The people inside were rescued along with their tents from some trees that luckily caught them a bit further down the hill.

In the morning, all ten of my toes were red, hot, swollen and tender, all sure signs that they were badly infected. Thankfully, the local guides knew what to do. They immediately cut some branches to make a stretcher, tied me to it, and then traipsed easily over the hills, carrying me, for the two whole days it took to get back to the airstrip. On one hand, I enjoyed not having to work so hard to climb the steep hills. But on the other hand, I was helpless, and with very infected toenails. And it was strange going down the steep hills and realising that if I wasn't tied to the stretcher, I would have slid off! The plane was called in to take me back.

Back in Lae again, I was admitted to the local hospital. The doctor made it quite clear that hippies were not welcome and then pulled out all ten toenails – without any local anaesthetic! It was an incredibly medieval experience and tremendously painful. At least the doctor was wearing surgical gloves and not dressed up like a character from the Spanish Inquisition! I imagined it was a bit like what torture would be like.

Afterwards they gave me some antibiotics, my toenails grew back almost perfectly in time, and I didn't die. I couldn't wear any shoes to start with as the exposed nail bed was supersensitive, so thongs continued to be my preferred footwear. But I never let my toenails get too long again.

Sing-sing and dope

Around the middle of my first year in New Guinea, the Goroka sing-sing was happening. The Highlands capital of Goroka was about 300 kilometres due west from Lae along the very rough Highlands Highway. A sing-sing is an annual gathering of many different villages or tribes, who are usually dressed in traditional regalia, to show off their culture via their dress, dance and music.

The Goroka sing-sing was set up by kiaps in the 1950s and has long been one of the most famous, with typically a hundred different tribal or village groups. A mate and I drove to see it in my trusty Land Rover. He used to drive semi-trailers along the Highlands Highway, a road that was essential to commerce in that part of New Guinea but was always collapsing or sliding away due to the heavy rain and its quick construction. The mountains were very steep, and the administration could not afford to build an all-weather sealed highway, so it was cheaper to just repair whatever slid away.

But my biggest worry was about accidentally running over a pig on the way there! Pigs were highly prized, and I was strongly advised by my kiap mates that if this worst-case scenario came to pass, under no circumstances should I stop but just keep driving away as fast as I could. In the heat of the moment, tensions would be high, and bad things would likely happen if I stopped. Instead, I should head to the next police station, explain the circumstances, and then leave behind whatever money the cops reckoned was fair compensation.

On the way to Goroka we had to stop for about four hours because the road had just slipped down the mountainside. While bulldozers very roughly cut a new road, my mate told me how on one occasion, after being stopped for about half a day, he drove over the freshly repaired section, looked down the steep mountainside and saw part of a semi-trailer buried in the 'foundations' of the dirt track he was driving on, the smokestack poking up at an angle. That road had collapsed under the semi due to big rain and it was just left there and built into the road!

At the sing-sing, thanks to carrying a movie camera and fancy tripod, I was able to talk my way onto the top of the covered audience stand. It was amazing to film the different cultures gathered together in one location: the local constabulary in a pipe band, the white-and-red makeup and long curved pig tusks through noses, incredibly ornate headdresses up to half a metre high and wide, fearsome weapons of war and shields, tortoiseshell armbands decorated with red string, groups dressed as raiding parties with terrifying masks and bodies covered in white ash, and of course, the local speciality, Asaro Mudmen covered in white clay with huge helmets also made from mud.

It was all new to me, and I used up rolls of film. It was impossible not to notice the only other long-haired white man, also in his early twenties, taking photos with a long lens. He looked at my Bolex movie camera, long hair and beard, and said, with an American accent, 'Hey, man, you look cool. Are you a head?'

I wasn't sure what a 'head' was, but I replied, 'Sure.' (I soon found out that 'head' was hippie jargon for a fellow dope smoker. The connotation was that smoking dope lifted you, and your head, to a higher plane of consciousness.)

Naturally the next thing he said was, 'Hey, man, you want some dope?' So again I said, 'Sure.' I had been wanting to try dope for ages, and despite trying for months to score in Wollongong, I'd never managed it! The deal was that I would give him money and he would send me the stuff in the mail. There was no offer to share a 'smoke' on the spot – his was a purely mail-order business!

So I gave him some money on the top of the stand at the Goroka showground, and about a week later, six little film canisters packed with marijuana appeared in the mail.

Dope and me

Of course, I couldn't ask my new mate on top of the Goroka stand for advice about how to actually smoke a joint – that would unmask me as a newbie. So back in Lae I bought some tobacco and papers and started off by teaching myself to smoke cigarettes. (Which I guess means marijuana was a gateway drug to tobacco for me!) Once I could smoke cigarettes without coughing my guts up, I started adding dope to the tobacco, increasing the amount I added each time. At first I didn't notice any effect, but then one magical night, everything clicked.

On that night, I unexpectedly found myself in the kitchen devouring pâté and thinking it was the most delicious food experience of my life! Pâté wasn't something I had ever eaten by the bucketful before that night while stoned. I listened to Jimi Hendrix through my headphones and for the first time felt like I was fully appreciating stereo music with the shift from one ear to the other. I felt like I got so much more out of the whole experience.

Of course, I would never recommend anybody break the law like I did – I am just going to blame my youth and inexperience. But there is an important caveat to smoking dope.

The thing I did not realise at the time was that dope destroys your discrimination, a fancy phrase that means you can't tell the difference between what is artistically good or bad. I recently watched a doco about The Beatles where they talked about going onto the roof of Abbey Road Studios (where they laid down some magnificent albums), smoking some dope and then coming downstairs into the studio to record. The music sounded great at the time, but replaying it the next day, they knew it was terrible. Demolishing a whole tub of rich pâté on my own clearly showed my lack of discrimination. I still have to say it was pretty damn tasty, though!

Paracelsus, the father of pharmacology, some five hundred years ago said something along the lines of, 'All drugs are poisons, what matters is the dose.' My father told me that when he enrolled at university in his teens, he started drinking very heavily. He stopped when he realised his memory was failing him. He reckoned it didn't get any worse once he stopped drinking regularly, but it never recovered to its previous excellent levels either. Through plain dumb luck, I got through lots of dope smoking with no real damage. My guess is that the luck was mostly that I started smoking in my twenties *after* my brain was fully developed. Now, as a boring grown-up, I think that having significant quantities of any drug (alcohol, dope, etc.) before your brain has matured can lead to permanent harm. In general, I would recommend not taking any drugs, especially before the brain is fully developed (say, twenty-five years of age), and even more so if that drug is illegal where you live. On the other hand, when have The Youth ever taken sensible advice? We have proof, in writing, that parents have complained about their kids for at least five thousand years. Almost certainly they've complained about them since time immemorial!

The human brain starts off with a lot of prewired firmware (how to breathe, automatic circuits for heart and gut, primitive vision, and so on). But stuff like language and social skills have to be learned along the way.

Speaking very roughly, one way to gauge the maturity of the brain is by sleep patterns. Immediately after birth, babies sleep a lot (not necessarily at times convenient to parents). The hours of sleep required drop closer to adult levels around puberty when, as part of all the other hormonal changes, you start producing more melatonin, but later at night. Melatonin can help you settle down to sleep – and then to sleep through the whole night. So making more melatonin, at a later time, means that teenagers probably need about ten hours of sleep (instead of the adult average of eight) but get tired later (say, 11 pm or after). So most teenagers spend their high-school years in some degree of sleep deprivation, because they can't get the ten hours' sleep they need. They fall asleep late but have to get up early for school. Teenagers are not evil when they want to stay up late – it's just hormones. (Mostly melatonin!) Once you get past your early twenties (again, speaking very roughly), you tend to lose the ability to stay awake effortlessly until 4 am. Bingo – this is a rough measure of your brain being (mostly) hardwired. Ta-da – now you're an adult.

I count myself lucky that my brief encounter with 'substances' was after my brain was mostly matured.

Cairns to Sydney – a close call

Come the Christmas holidays I headed back to Australia. Just for fun, I decided to fly from Lae to Cairns, and then train or hitchhike some 2500 kilometres down to Sydney. The trip ended up totally out of control – but luckily, I survived.

I started at the Cairns waterfront, but I hadn't realised it was mangroves, not pristine sand. (Back then I did not at all appreciate how important mangroves are in supporting biodiversity, being a fish nursery and protecting the coastline. I just saw them as mud.) I was imagining sparkling yellow sand beaches, and this was a letdown.

I quickly gravitated to the local 'heads' to improve my experience. A friendly local and his girlfriend took me around Cairns. We smoked some dope, and they invited me to swim with them across a small lake, but with my head all fuzzy, I completely forgot that I was a very weak swimmer. I got about 50 metres across the lake before I was exhausted. Luckily for me, they were excellent swimmers, and they calmed me down and managed to swim me to safety. That was a close call.

Then things just got messier and messier!

As part of the holiday, I wanted to take the train south to Townsville. I planned to get some dope for the trip but had no luck. Some of the 'heads' told me their back-up high was to take a packet of antihistamines with a packet of non-steroidal anti-inflammatory drugs. I didn't ask for any more details, and it sure didn't turn out as I hoped.

The train took all night to travel a lousy 350 kilometres. (If only we had fast trains in Oz, though if we did then maybe I wouldn't have been able to get off later!) I had boarded the train in the late afternoon, foolishly swallowing both packets of tablets and giving no thought to potential side effects. I didn't even realise that non-steroidal medications could give you stomach ulcers and bleeding from the gut, or that antihistamines in high doses could cause drowsiness, agitation and confusion. I did get really intoxicated, but it was no fun – for starters I had gaps in my memory, and also had massive muscle cramps in my arms. In retrospect, I was really stupid to take the medications like that!

The carriage had a long skinny corridor down one side and a bunch of compartments with facing bench seats on the other. Afternoon turned into night, and random fellow travellers flickered in and out of my troubled consciousness. I vaguely remember walking up and down the corridor visiting people in the other compartments, and I'm pretty sure they weren't thrilled by my company! I had intense conversations with several different people while having to stop to bend and straighten my arms continuously.

My next memory was of lying comfortably on a bench seat in the daylight. I felt normal and unexpectedly well rested. Very cautiously, I opened one eye a fraction. Oh no – there was a Queensland cop sitting directly opposite me! He was wearing the old drab olive uniform and was armed with a gun, strapped in its holster. I immediately assumed that my drug-crazed behaviour the previous night had annoyed my fellow passengers, and somehow the cops had got involved.

I lay very still, breathing quietly and trying to work out a plan. After about half an hour he got up, I guess to go to the toilet. Through a single narrowed eyelid I saw him turn right into the corridor. I immediately grabbed my backpack, turned left and ran to the other end of the corridor, opened the door and, without thinking at all, jumped straight off the train!

Luckily the train was moving slowly. Jumping off a moving train was another very stupid idea (guess what – don't take drugs!), but amazingly, I was not hurt (much). I jumped the railway fence, ran to the road, stuck my thumb out and a car stopped almost immediately. Desperate to get away from the train, I blurted out, 'Sorry, I've just had a family emergency and have to get to Sydney. Where's the nearest airport?'

Once again, luck was with me.

The reason the train was slowing down was because we were coming into Townsville. It had an airport that did single-hop flights directly to Sydney, which meant I could avoid landing in Brisbane (just in case the Queensland cops had rung ahead to bust me). The driver went out of their way to take me all the way to the airport. I arrived with enough cash in my wallet to buy a ticket and perfectly on time to catch a direct plane to Sydney. I ran to the gate, boarded the plane and finally breathed out as we lifted off.

Things could easily have turned nasty. I could have been arrested by the cop and my life would have been headed down a very different road. But ridiculously good luck saved the day and me.

Science in Sydney – with hipsters

A few days later, with those Queensland narrow escapes behind me, I headed to the CSIRO's Division of Textile Physics in Sydney. My pre-arranged mission was to learn everything I could about hair and wool in two short weeks. The CSIRO division's focus was mostly on wool, which is much more elastic than human hair, but there are some similarities, so to my boss my placement there was still time well spent.

His plan was for me to spend a few weeks of my holidays boning up on the properties of hair and wool on his behalf, because he was still trying to get me to be part of his dodgy PhD. I stuck doggedly to doing my actual science and taking real measurements, hoping I could get away with that. The idea was that I summarise everything I learned at CSIRO in a document for him. I still had no intention of helping him fudge his PhD, and I didn't really want to spend my own holidays doing research on his behalf, but on the other hand, I was genuinely interested in the topic.

Strictly speaking, of course, my boss shouldn't have expected me to do any work in my holidays – let alone *his* work! But it really turned out well for me in the end. I had a great time at CSIRO and made so many new friends. As is often the way with me, my habit of just going with the flow ended up washing me into another interesting phase of life.

But back to wool, which can be classified in many ways – colour, strength, diameter and so on. Back then, the only way to measure the diameter of a wool fibre was to look at it under a calibrated microscope, taking measurements slowly along the entire length. You could easily spend half an hour taking thirty measurements and then averaging them, and that was just for one single wool fibre. There were many thousands of fibres in a kilogram of wool, so this was a painstaking process!

One of the CSIRO scientists who seemed extremely cool (he had very long hair and a beard) was also really smart. With some impressive lateral thinking, he'd worked out a way to use a newfangled tool called a laser to measure all the fibres in just a few minutes. The light from the laser was bent differently by the fibres, depending on their diameter, creating spots of light and dark. Cleverly, he used another newfangled tool (a computer) to tease the information he wanted out of the 'photos' of these interference patterns of light and dark patches. Wow! These were huge breakthroughs. He was like a genius.

Wool finer than 25 microns (a micron is a millionth of a metre, and human hair is about 70 microns in diameter) is considered suitable for clothing. In 2004, a bale of the then finest wool produced (about 11.8 microns) sold for A$3000 per kilogram.

After a few days of working with this brainy, cool CSIRO scientist, he invited me around to his place in inner Sydney (so hip!) for dinner. I was excited. He lived in a share house with musicians who were in a rock 'n' roll band and knew interesting people across all the art scenes.

As soon as I walked in he offered me a big fat joint, much more powerful than anything I'd had in New Guinea. By the time I woke up the next morning, I had had my first taste of sex, drugs and rock 'n' roll all mixed together, and I liked it a lot! Needless to say, I spent as much time as I could with this guy and his friends while on secondment to the CSIRO.

When I flew back to New Guinea for the academic year a few weeks later, I felt like a changed person, less insecure and inexperienced. I knew now that there were little pockets of community where I fitted in perfectly, which was a nice experience. I'd had a taste of fun and I wanted more!

I continued my work with the students. I loved lecturing and absolutely loved the tutorials where I could work one-on-one with them and watch the flash of enlightenment when they truly understood a new concept for the first time.

Unfortunately, I was also having to slog away and do all the grunt work for my boss's PhD. It was soul-destroying because he didn't even want to see the real results or the genuinely interesting data I was collecting. And because of this, about halfway through my second year at the HiTech, I was really unsettled and looking at job ads in the newspaper.

A precursor to the Australian Film and Television School was offering scholarships. I applied for the very first intake by filling out a form – they didn't even ask for any examples of my work at all. To my absolute surprise, I got a scholarship.

The scholarship was my escape route out of the job in Lae. I resigned (so technically, yep, I kind of 'failed' at my second professional job – in that I felt pushed out before I was ready) and headed to Sydney to start my new and hopefully avant-garde life as an underground filmmaker.

1971.5–75

Budding filmmaker

Around mid-1971 I arrived back in Sydney, full of enthusiasm. I had some money saved up, a 16-millimetre movie camera, and an arrangement with the incipient Film and Television School (FTS) to cover the costs of film and editing of the Great Australian Movie I intended to make.

I had three really good reasons for wanting to get into filmmaking as a career. First, I loved watching films. Second, films were a perfect blend of function and form (in other words, a mix of science and art). Third, making films would be another clean break with my recent past, further cutting any ties with my two foiled careers in science.

I was going to be an artist now, not a scientist! Luckily I still had my mental toolbox to let me jump across into another completely new area. I was hoping that adding 'art' to my working life might show me a different way to find success, happiness and fulfilment.

It didn't turn out that way. I didn't reach the heights of the first FTS graduates, such as Gillian Armstrong (director of *My Brilliant Career*, which starred a young Judy Davis, and many other successful movies) and Phillip Noyce (director of *Rabbit-Proof Fence*, *Dead Calm* and plenty more). My slim list of credits, after several years, amounted to getting one film onto a music video TV show and, quite separately, helping set up Australia's first cable TV network.

In retrospect, despite having the best of intentions, I reckon the main thing stopping me from making movies was quite simply me – and very likely my marijuana-soaked lifestyle was the biggest culprit in terms of blame!

Certainly, the FTS tried to help. Along the way they sent me many letters inviting me to talk with them. That was a very reasonable request, seeing as they were paying for my film stock and, later, the hire of the editing suite. Back then I didn't realise the importance of communication and personally meeting with people who were keen to help. Maybe I could have become a famous Hollywood movie director if I had even taken the smallest amount of advice from the FTS!

Ironically, my little detour into being an 'artist' gave me a road back into a career in science. Filmmaking turned out to be essential background training for my later role as a hospital scientific officer.

Underground movies

I went to weekly underground film nights in Sydney that showcased up-and-coming filmmakers. These films were mostly three to five minutes long, which maybe explains why I hate sitting through today's three-hour epics.

One short was called *Bolero*, named after Ravel's musical piece from 1928, which was also the soundtrack. Ravel's *Bolero* has a very simple and compelling quality: the theme is repeated over some fifteen minutes, but with added instruments, and always getting louder. The subject of the film *Bolero* was a woman sitting on a chair wearing a stereotypically Spanish outfit, with a rose held between her teeth. She stayed perfectly still while, over some ten minutes, the camera slowly and steadily moved towards her. The shot very gradually changed from her whole body filling about a tenth of the frame to the rose between her teeth filling the whole frame. It blew my mind, though now it seems hard to explain why. It was probably largely due to my mind already being blown by 'substances'!

First go at filmmaking

Based on vague concepts about peace and love, I came up with an idea for a three-minute movie called *War Chant*. At no stage was there a coherent script.

I had recently read a book that came as a bunch of individual unnumbered pages in a box, and it influenced me hugely. It seemed so modern and ahead of the curve to not even have a fixed plot for the story – absolutely everything was up for grabs. It was bizarre and crazy and never made any sense, just like my film! Instead of a coherent script, my 'vision' was to have a whole bunch of special effects layered on top of each other.

After a long period of doing anything but the job at hand, I tried to get my act together. Finally I decided that the actors in my masterpiece would run across a natural Australian bushland in front of the camera. Over and over, they would run from left to right across the background. I would film this with a variety of lenses (wide-angle to telephoto) and with a variety of film stocks (colour, black and white, infrared). Then I'd throw in a few random close-ups of faces flickering with intense emotions (*War and Peace* Russian angst kind of thing). Then, easy-peasy, I would edit all these clips together, overlay some music from the band I was a roadie for (more on that later) and end up with a three-minute extravaganza!

I organised a loan of some military uniforms to fit with the war theme and got the Ektachrome film stock, some tents for the overnight stay, and the actors. And then, with everybody in my Kombi van, we rolled into the Royal National Park just south of Sydney with a key to get in through a locked gate and permission to film.

Needless to say, I really didn't know what I was doing. The very kind (and unpaid) amateurs ran at various speeds across the landscape, holding toy rifles. We even had a small child on hand to convey something that I couldn't verbalise about the horrors of war, but I hoped that it would all make sense in the edit.

We started on Saturday afternoon, stayed overnight, and left on Sunday afternoon. Interestingly, halfway through the filming, a police car came in, parked, watched me filming people with toy rifles running around, and then drove away again without a word!

The editing stage should have been ridiculously easy, but sadly, that's where everything came undone. It became blindingly clear to me that the lack of coherent script really made everything else pointless. I tried different cuts, but I could never get what I wanted. I was heartbroken. My dream had turned to dust!

My coping strategy was a mix of avoidance combined with denial. I never showed up to any classes at the FTS and never responded to any of the requests from the lecturers. I just let my Big Chance slip away.

I didn't disconnect from film altogether, though. I still loved going to the movies, and I read Hollywood technical journals such as *American Cinematographer* every month, but I studiously ignored the letters from the FTS. My dream of making films was fading, but I hung on to the film camera, for no good reason.

Second go at filmmaking

About a year later, one of my mates came to me with an offer. He was managing one of the early Australian heavy-metal pub-rock bands and asked me to come to one of their gigs. 'Shoot a live music video of the band and we'll give you some money.'

A chance at salvation! No thought required, I just said yes. It sounded like a simple job with a beginning, a middle and an end. I could do a reconnaissance at one of their live gigs, work out what was needed, turn up on another night to shoot it, then edit it and deliver.

Of course, there were many problems. Of course, I should have asked more about what the gig would involve right from the start. And of course, I should have asked how much I would get paid!

At the reconnaissance gig, I realised the venue was so poorly lit that even the fastest then-available film (160 ISO) couldn't capture a usable image on a movie camera. A higher ISO allows you to work in lower light levels, but the image is lower quality. With my movie camera, each exposure was fixed at one-fiftieth of a second. With a still camera, the shutter can be left open for much longer to let more light in, so long as your subject didn't move. But movie cameras don't have that option – they must expose twenty-five frames each second to avoid jittering.

Luckily, this was a technical problem that I could handle. I simply had to work out how to make the film 'faster', or more sensitive. I wanted to take the film speed up from 160 ISO to 1600 ISO (ten times faster).

I went to the head office of Kodak (former number-one film supplier on the planet) in the Sydney CBD to study their vast array of technical books. After a few days in their library, I found a workaround that would involve some special processing, but Kodak would not do it for just a few rolls of film. I reckoned I had enough chemistry, physics and general problem-solving skills to do it myself!

So one night in the laundry at home, I tried to follow the instructions in the old Kodak technical manual to process some film at 1600 ISO. The results were terrible.

First, some of the chemicals were toxic – I felt dizzy from the smells. Second, there were thirteen separate steps in the process, and most of the different liquids and chemicals in the processing had to be kept within 0.5 degrees of 29 degrees Celsius, although some gave the luxurious wriggle room of a few degrees either side. Even with a lot of preparation and care, the temperatures of several chemicals were out of range by several degrees, which meant the colours were all wrong and there were splotches on the film. Bummer!

I rang around my circle of techy friends for help. My physicist mate from the CSIRO designed and built me a working prototype of a brand-new technology temperature control circuit (proportional zero-crossing, if you must know). It worked beautifully. A second mate very skilled in electronics turned this prototype into functioning units. And a third friend, Jackie Joy, helped me integrate the temperature controllers into a mini colour-processing laboratory in my laundry.

The various chemicals were mixed up fresh in their own bottles and then brought to the correct temperature by sitting the bottles in a bathtub full of temperature-controlled water. (Forget a tolerance of 0.5 degrees Celsius – we were getting 0.1 degrees! Even the Kodak people were impressed when I chatted with them about my progress.) One at a time, the liquids were poured from their individual containers into the processing drum that had been pre-loaded with a hundred feet of exposed Ektachrome film. This little processing drum also sat in its own separate temperature-controlled bath. After an hour of pouring liquids back and forth in the right order, the film was ready to be removed from the processing drum and hung to dry in a cabinet, with filtered air blown

A Periodic Tale

through. After a few hours it would be dry and ready for editing.

Building all of this machinery took several months. And I still had to earn money to live on by driving taxis (more on this later), which took several days out of my week. But I was very encouraged in that test run, when my film-processing lab worked perfectly.

I went to the next gig the band played, shot some film, then processed and edited the film and matched it to the backing soundtrack. It was 100 per cent successful and ended up a nice little music video, if I do say so myself. In the end, the processing of the film turned out to be the shortest part of the job.

I went to deliver my rock 'n' roll movie, feeling on top of the world. The band was staying at what was then *the* hotel for musicians, the Sebel Townhouse in Kings Cross. Staff famously turned a blind eye to the excesses and bizarre requests and behaviour of their guests, who included Elton John, Dire Straits, David Bowie, Rod Stewart, the Harlem Globetrotters and more. It was so famous that it featured in *Abba: The Movie*!

When I got to their room, the band were drinking champagne and smoking very high-quality Buddha stick. I said, 'Well, here's the film – it turned out great.' I can't remember the exact words, but the response was along the lines of, 'Here's forty dollars. Now f**k off.' That didn't even cover the cost of one roll of film!

I decided there and then that there was no real future in making music videos and movies. A monumental error on my part! (Up there with those who claimed there was no role for personal computers in the home.) The cable channel Music Television (or MTV) took off like a rocket in 1981 and launched many careers in film. But not mine!

Australia's first cable TV network

My film career took an unexpected detour thanks to a mate who was part of Bush Video, a collective of artists and technicians using the new technology of portable video recorders to film gigs. They got some government funding with virtually no strings attached. It allowed Bush Video to do low-cost coverage of anything, anywhere, and with a very alternative bent. In 1970s Sydney, I did some work with Bush Video as a camera operator.

The organisers of the 1973 Aquarius Festival, a music and countercultural arts event held in the tiny dairy and banana town of Nimbin near the Queensland border, wanted Bush Video to provide festival participants with video access to record whatever they liked and play it back on a network of TVs across the site. This concept of 'narrowcasting' was a very radical idea then. Up to that time, 'broadcasting' was how all media worked, where a small number of people controlled the content and then 'broadcast' it out to a wide audience. Bush Video went down a completely different pathway where not only was the audience small and directly targeted, they could also respond to what they saw, and generate their own video content.

My Bush Video mate knew I was coming to Nimbin

anyway as a roadie with a band booked to play (more roadie stories later). He rang me, saying, 'Karl, we're making all this electronic video stuff and need about a kilometre of video cable brought up. Can you do it?' I was happy to help. I'd swapped the Kombi for a 1954 Chevrolet that could pull 140 kilometres per hour and had a huge boot.

I turned up to the address he gave me, where a forklift truck deposited a roll of cable about one and a half metres in diameter and half a metre high into the boot of my Chevy. It weighed several hundred kilograms, and the back sank down as the rear leaf springs turned inside out. (My heart sank in tandem with the springs. The springs never recovered, and I later had to remove them and get them reshaped and retempered.)

Once my band's gig was over at the festival, my roadie work was done, leaving me with nothing but free time. So I got totally involved with the tech side of Bush Video.

They had fancy black-and-white video cameras that were very advanced technology for the time. The video recording heads were tiny (smaller than the head of a match) and spun at very high speeds over the slow-moving magnetic tape to capture the data fast enough for movies. Unfortunately, they were also temperamental and would scrape magnetic 'dust' off the tape, which after a short time would make the image go bad. So then we'd have to delicately wash the tiny recording heads with isopropyl alcohol to remove the magnetic dust and bring the video recorders back into operation. This would work for an hour or so, and then we had to clean them again.

We lay down the kilometre of video cable (and a kilometre turned out to be only barely enough!) and set up video cameras all through the town of Nimbin – in people's houses, in shops, on the pavement, in the fields where people camped, in their tents, up trees … We also had to wire-in TV sets wherever there was a camera station. It was wild!

Of course, all the video cameras and TV sets in the field had to be wired through our control room, which had little black-and-white monitors and a few video cameras. None of this network had been formally planned, which meant we had to design and then wire up everything on the spot. It took days to complete, and I was often soldering until 3 am. We ultimately made Australia's first local cable TV network, physically linked by the kilometre of video cable and us, a bunch of electronic hippies in Nimbin!

The Aquarius Festival was an amazing counterculture experience in beautiful bushland. And it was much more. The really impressive thing about this was not only did we build the first streaming cable TV network in Australia – it was *two-way* video streaming, where the viewer could also be the creator. Wow!

1971.5–75

Roadie

Life is full of surprises.

I arrived back in Sydney from New Guinea feeling like I had all the time in the world. My plan was to live off my savings and the small amount of funding I'd been given to make movies, and become a successful underground filmmaker. I had no idea that it wouldn't pan out that way.

On landing in Sydney, the first thing I did was head over to my CSIRO mate's place. As luck would have it, he had a spare room to rent! I promptly moved in, and one of the other housemates was a drummer who was working hard with a bass guitarist to kick off their rock 'n' roll careers. They listened to all kinds of music, including blues, which I didn't really understand at the time. It was a great education for me. Before then, I didn't realise that American blues is the basis of rock.

They set up a classic three-piece rock 'n' roll band – drummer, lead guitarist and bass – and started practising at our place. The band's name changed frequently, but Wasted Daze was my favourite. They all took turns being the singer. They played a lot of covers, and they also wrote their own songs. I found smoking dope and listening to music with the band a lot more attractive than filmmaking, as it turned out.

I quickly fell into becoming a roadie – it was stupidly easy. They were a good band and quickly built up a dedicated fan base, but they were still small fry. The band got to a point where they could hire a local space like the Paddington or Balmain town halls and charge for entry and make a bit of money. I morphed from listening to them play at home, to going out with them on gigs and helping them set up, as one of just two roadies.

Roadies

Roadies are part of an artist's travelling support crew, and they pretty much do *everything*, bar perform on stage! The size of the crew depends on the size of the band. There might be different technicians (or techs) who set up and tune the instruments on stage beforehand, adjusting the audio levels. There might be front-of-house and monitor engineers, lighting designers and techs, as well as the various managers (tour, production, stage) plus security, truck drivers, merch crew, pyrotechnicians and more.

One of the band members was an electronics whiz and built all their amps from scratch. It was slightly illegal, in the sense that it was totally illegal. He had a day job at a major electronics wholesaler and retailer, and every day he would 'liberate' some modest number of parts and then assemble them into powerful 100-watt (which was huge back then) amplifiers.

As a de facto roadie, I would load my Kombi van with their audio and lighting gear, unload and set everything up on stage, including running audio cables to the mixing desk. I quickly learned that 'bump in' meant taking everything out of the truck and setting it up, while 'bump out' was the opposite. I still regret not learning how to operate an audio mixing desk back when I had lots of spare time. I also regret not learning how to play a musical instrument.

The band performed at all sorts of places – high-school formals, dances, private parties and music festivals. After bumping in, I would dance wildly to the music with the audience, having a ball. These were literally my youthful days of sex and drugs and rock 'n' roll.

One strong memory that I have with the band was at a spontaneous big party for a few hundred hippies, with a few bands performing. It was held in a natural amphitheatre in Ourimbah on the Central Coast of New South Wales. Everyone was smoking dope and grooving away. But halfway through the night, some uniformed cops started slowly circling around on the ridge above the valley. I was rather paranoid that I would get busted, and really thought that we should dump the dope. But my fellow heads told me it was all cool and not to worry – and the bands kept playing, the audience kept dancing and smoking dope, and the cops kept circling. After a few hours, the cops went away. I think that I remember this so clearly, because it brought home to me how my new identity as a hippie simultaneously gave me a heartland group (with whom I felt very relaxed) but made me feel a little paranoid and on the edges of society.

My attitude about the police has shifted over the decades. From what I now know, policing is a very hard (but essential) job at the best of times. Over the years I have worked with police in hospitals and other roles, and most officers are decent individuals trying to make a difference. The gruesome and violent things that police have to deal with would deeply shock anyone – even people who also are up to their elbows in blood and trauma, such as emergency hospital workers. Without a supportive culture inside the police force, it's hard to imagine how they could survive their day-to-day work.

Killer Kombi

At the time, the lead guitarist and I both had a VW Kombi, so between the two vans there was enough room to shift all the bulky gear. Kombis had volume, but they lacked both power and safety. Which was obviously a poor compromise!

On one occasion, the guitarist was driving back from a gig in his fully loaded Kombi, plus two passengers in the front on the bench seat. Behind them was a very thin metal barrier (euphemistically called a 'safety barrier') that came up to mid-chest height – and behind that was all the equipment.

As always the van was stuffed to the gills. Suddenly he had to hit the brakes hard, causing the massive load in the back to slide forward just a few centimetres. The so-called safety barrier buckled that same number of centimetres, creeping up on him like a wall in *Indiana Jones*. It turned out that the safety barrier had the structural integrity of a cigarette packet, and if there had been a real collision, they could have died! So shortly afterwards, the band bought a proper two-tonne truck with a cab and a self-contained back section for the equipment. It was better in every way, and so much safer.

After a few years, the band was looking and sounding really good. In 1973 I heard about the Aquarius Festival coming up in Nimbin. The festival was promoted as our Australian mini Woodstock. With an expected audience of five to ten thousand people, it was minuscule compared to the 450,000-strong audience at Woodstock, but it was a good crowd for a gig in rural Australia back then!

I really thought this could be a breakthrough gig for the band, so I convinced them to drive the solid 800 kilometres or so from Sydney to Nimbin to perform for free. On the one hand, they would have to pay for petrol, but on the other hand, it was a chance to get 'discovered' and become fabulously wealthy and famous!

Trying for the big time

The band did a fantastic show to a very enthusiastic audience, but missed out on a Magic Break to Fame and Fortune at Nimbin. Bummer. Magic breaks were tricky to come by –

magic mushrooms, on the other hand, were pretty easy to find up there!

The band tried everything they could to get famous. Once they lobbed into the studio of the then top Sydney AM radio station and did their best to convince the big-name disc jockey of how good they were, asking him to pretty-please play their music on his show. The jock replied, 'If you've got cash for my pocket and my managers', then you've got a deal.' The guys couldn't even scrape a lousy fifty bucks together between them, let alone the several hundred more he was asking for a paid promo! So the direct approach was as unsuccessful as playing the Nimbin festival in terms of lucky breaks.

One thing I learned later about the music industry was the three-year rule. It says that, on average, an up-and-coming band has a lifespan of three years in which to crack the big time – or otherwise they just fade away. The band was good, but they never got to be famous.

Bo Diddley

One of the unexpected highlights of this time for me was working as a roadie for the blues icon Bo Diddley.

We were about two years into that three-year window of trying-to-be-big when, out of the blue, the band got a call from the agent for Bo Diddley. Bo was a major force in blues in the United States with songs like 'I'm a Man' and 'Who Do You Love?' He influenced many musicians, including Elvis Presley, The Beatles and The Rolling Stones.

Bo was coming to Australia for a one-month tour, and his agent offered the band the job of backing him on the Sydney leg. This really was a lucky break – it seemed the band's love of blues had finally paid off. Hoorah!

Bo arrived and started practising his set with the band. It didn't take long for them to get the 'sound' that he wanted. As the shows started, something began to bother me about his having to perform the same songs night after night. While we were bumping out one night, I asked Bo about it. 'I've heard you play the same songs a bunch of times now, and each time you play with the same enthusiasm as if it's the very first time. How do you do it?'

He put a huge hand on my shoulder. 'How come showbiz is called showbiz? Not show fun or show play, but show*biz*. The biz is that they give us money and we give them a good time. So for me, it might be the four-hundredth time I've played that song, but for them it's their first time, so I'm gonna play it like it's the best thing they've ever heard. And then they're happy, and I'm happy, and we're all happy. What's not to like?' I loved that lesson in work ethic and performance, and I've tried to incorporate that approach into every part of my life since.

A few decades later, the Malaysian government invited me to speak to school and university students in Sarawak – with the main event having an expected audience of about a thousand people. When I turned up, there was one solitary person in an otherwise empty theatre! Turned out there was a multitude of reasons for the lack of audience: there had been an earthquake, an unexpected tropical monsoon had lashed the island, roads were blocked, it was university holidays instead of university term, the person organising it had left, and so on. Like a beacon in the night, Bo's words echoed in my head and I gave the best goddamn power-packed show to that one person. I felt like a true professional, and the Instagram photo of the huge hall with just one member sitting in the audience did really well!

When Bo toured the next year, he got us back as his band. I loved the vibe so much that I would roadie for Bo with the band even after I started a Proper Job again as a hospital scientific officer.

The arts

I had other friends who worked in the music industry, and one time I was given complimentary tickets to see Tina Turner perform – and to the after party. I even got to dance with her in her palatial suite at her fancy hotel!

Years later, I heard a story about her from an audio tech.

When Tina was scheduled to perform her anthem 'The Best' at the 1993 rugby league grand final in Sydney, the audio team installed the usual huge array of speakers, so everybody in the stadium could hear her singing. As usual, they also installed foldback speakers pointing at the stage, so she could hear herself and the band. Apparently, Tina was fairly deaf (very common in the music industry from all the exposure to loud noise) so they put in a *lot* of foldback speakers and really cranked their volume to max. But during the performance, the audio techs somehow forgot to turn on the audience speakers.

Luckily, nobody in the crowd noticed because the volume from her foldback speakers alone was enough to easily fill the stadium! I don't know if it's true or not, but it's a funny story.

A little bit of dope

As part of my wild, crazy, hippie, roadie lifestyle, I found myself smoking marijuana every day and enjoying the experience immensely. There was a period of about a year when I would roll myself a joint before I even got out of bed. My unofficial goal was to smoke my own body weight in dope. Crazy!

Smoking so much dope made me very gullible (dunno if this happens to everyone). I read the 1961 sci-fi novel *Stranger in a Strange Land* and was convinced that, with enough training and practice, I could move objects with my mind! I was very naïve and impressionable. But then, so were the CIA and the US military, who sunk millions of dollars into psychic research! Watch the funny send-up movie *The Men Who Stare at Goats* and laugh at their craziness.

While I consumed huge quantities of dope, I took very little acid in the 1970s, although it was big at the time. My knowledge about acid trips came mostly from reading widely, especially the books of Carlos Castaneda, starting with *The Teachings of Don Juan: A Yaqui Way of Knowledge*. Carlos purported to have apprenticed himself to a very wise Mexican shaman named Don Juan, who taught Carlos the power and pitfalls of psychedelic drugs. In the books, Don Juan flew his spirit to distant locations, where he not only observed but even took part in activities without his physical body. It later turned out that the books were works of fiction based very loosely on Castaneda's studies as an anthropology student at the University of California. But they were useful for me in teaching a respect for the power of psychedelic drugs – and that they were not something to be taken lightly. Indeed, today psychedelic drugs are being explored for potential use in some medical and psychiatric conditions.

While I was lucky with my small number of experiences with psychedelics, I knew plenty of people who had bad trips and got themselves into very dangerous situations. It can happen.

Rock 'n' roll survivors like Keith Richards (The Rolling Stones) and Anthony Kiedis (Red Hot Chili Peppers), to name just two, are different from the rest of us. Somehow they can consume huge quantities of drugs, keep crazy hours, eat nothing even remotely like a vegetable and get up in the morning with surprisingly fresh skin. My crazy conjecture is that they sweat a lot on stage, which helps moisturise their skin, so they stay looking good!

Tsunami madness – brown rice and kerosene

Another time someone in my circle told me about a prophecy that, on the next weekend, a giant tidal wave was coming from the Antarctic (as a result of massive glacial collapse), sweeping across Adelaide and then coming inland, only to be finally stopped by the Great Dividing Range, so the only safe place to be was high up. This was before Science and Logic ruled my world, so I totally believed them, no questions asked!

Looking for high ground, I headed up to the Blue Mountains with my girlfriend at the time and the band's drummer, laughably armed with the essentials to survive the imminent destruction of the southern half of Australia. We took my Kombi with a full tank of fuel, a 20-litre drum of kerosene (for our lamp and little stove), a 20-litre drum of water, cooking utensils and a 20-kilogram bag of brown rice. Of course, we also had 'essential' packets of tobacco, cigarette papers and a lot of dope.

We parked near Sublime Point Lookout at Katoomba to watch the catastrophe unfold while smoking dope. 'Surprisingly', the tsunami never arrived. We slept in the Kombi in the parking lot and drove back the next day, very happy not to be obliterated. I was weirdly disassociated from my lack of insight and judgement.

But slowly, I came back from the Dark Side, and became more rational.

In 1962, a report by the British Royal College of Physicians, *Smoking and Health*, was one of the earliest official acknowledgments of the health risks caused by smoking. In 1964, the US Surgeon General's own report was also quite clear about the dangers of tobacco smoking. But in 1973, aged twenty-five, I was not aware of either of these reports – and I was smoking a lot more tobacco than dope.

A muso mate of mine told me he had stopped smoking tobacco. A little surprised, I asked why. He replied that it was well known that smoking was bad for one's health, and that both the UK and US governments had released reports on this.

Wow, really? I checked out both reports at the local library and soon headed down the pathway of giving up smoking tobacco too. This time, I was lucky to be in a receptive state of mind to listen to rational advice. It wasn't too hard to stop smoking tobacco for me, and I certainly didn't want any long-term lung disease. However, I wasn't quite ready to give up the Sacred Herb, despite that being the logical follow-on.

Being a roadie was a great education in music and life. While my savings lasted, it was like a permanent holiday. But my savings ran out and I needed to make money elsewhere to fund my hippie lifestyle.

I didn't want to go back to working as a scientist, as there was still a bad taste in my mouth from those jobs at the steelworks and HiTech. Plus, I wanted to continue to keep my own hours and feel like my own boss.

And I definitely wasn't ready to cut my hair!

That's how taxi driving stepped in as my next sidetrack.

1971.5–75

Knight of the Road

I was now in my early twenties, bearded with gorgeous long hair all the way down to my bum, and I was driving taxis in Sydney for a living. I thought that as a taxi-driving filmmaker–roadie–hippie I was pretty cool. (Yeah, okay, I had gone off the rails and was definitely deluding myself.)

I only drove taxis on the big-money nights, Thursday to Sunday. Come the Sunday shift, I would do a fifteen-hour shift, starting at midday and finishing around 3 am on Monday morning.

Taxi driving was actually really good for me. For starters, it helped me get over my natural shyness. The night shift made it especially easy to settle into the chitchat that other people seemed to find so simple. I didn't need to make eye contact because of the dark, and anyway, I had to keep my eyes on the road! Being freed up of the need for eye contact let me relax, open up and learn to talk freely.

One time, a couple got in the back without breaking their conversation, as if I wasn't there. I didn't mind. But by listening and not talking, I picked up something odd about the way they spoke to each other. At a natural break in their conversation, I interjected from the front seat, 'It seems to me that the two of you are not really suited to each other – you're not compatible.'

The guy said to me, 'You're crazy – we love each other.'

But she responded almost simultaneously, 'Actually, Robert, I've been wanting to break up with you for a while.'

I wasn't surprised when Robert didn't give me a tip, but maybe I saved him from a lot of pain down the track! I've had a lot of different jobs, but sensitive couples-counselling was never destined to be one of them …

I could never have talked so openly to anyone, especially a stranger, before I was driving taxis!

Speed demon

As an uncertified speed demon, my best time from central Sydney to Parramatta was fifteen minutes, and from Manly to Sydney CBD was ten minutes. Mind you, both were at about 2 am, on totally empty roads and before fixed speed cameras. One of my taxi mates bettered my personal record by doing Manly to the city centre in just eight minutes!

I know this was pretty irresponsible and that I was working as a taxi driver, not a race-car driver. But on the other hand, how much money I could make was directly related to how many kilometres I could cover. In my serious taxi-driving years – from 1971 to about 1975 – I would regularly drive 400 kilometres each night. Most of my driving buddies averaged closer to 320 kilometres.

As a cabbie, I saw myself as a Prince of the Night, with the ability to go absolutely anywhere (so long as I was behind the taxi wheel). On one occasion in the early hours of the morning, I saw two cops unloading various valuable items from a broken pawn shop window into their paddy wagon. Perhaps they were securing the contents for the owner? I parked my taxi behind the wagon (in a no-standing zone!), walked up to the cops and asked them what was happening. They sized me up immediately as no threat, just a long-haired hippie taxi driver, and naturally told me to get lost. But they didn't book me or put the boot into me. To some extent, taxi drivers and cops were fellow Knights of the Road, and mostly we tolerated each other. This was strangely incongruous with my hippie paranoia about The Man when I wasn't driving a taxi. The taxi made me feel relatively untouchable.

More than once, police and cabbies worked together like something out of a movie.

One of my fellow taxi drivers told me that he was driving around Sydney when he suddenly found himself following his own personal vehicle. What the … Somebody had stolen his car!

He called in to the radio base, who immediately called the cops and also put the word out on the two-way taxi radio, which resulted in an entire phalanx of taxis heading towards him.

The police very professionally blocked the road in front of his stolen car, and the cavalry of taxis blocked any other possible exit. And *voila*, a very surprised thief was apprehended and one stolen car returned directly to its rightful owner.

It was a beautiful piece of symmetry and justice.

Strange rules

I read my way through the entire New South Wales Department of Main Roads rulebook while waiting at cab ranks rather than just sit idly. I found a rule that said that you are not allowed to carry a drunk or dead person in your taxi, except for the express purpose of taking them to the nearest police station.

This raised a lot of questions with my fellow taxi drivers, who were a philosophical and free-thinking group. I mean, taxis were *always* taking drunk people home, and most of these passengers were not asking to be driven to the local lock-up. And as if the police wanted all of Sydney's taxis shuttling drunk people to them all night – they already have enough drunks to deal with!

It's funny to think of that rule now, half a century later, given taxis are actively promoted as a 'Plan B' when you've had too much alcohol to drive yourself home.

Asleep at the wheel

Falling asleep at the wheel was an occupational hazard for taxi drivers given the long hours. Usually the shuteye was just a brief moment, but microsleeps can be really dangerous. If you drop into a microsleep for, say, six seconds while travelling at 60 kilometres an hour, you can easily cover 100 metres, and all sorts of things can happen in that time.

Early on in my taxi-driving days, I made the conscious decision to follow all the road rules every time I drove. There were no exceptions (apart from maybe going over the speed limit a *teeny-weeny bit* every now and then).

At every red stop sign, I came to a complete halt. I never just touched the brakes to slow down then kept on rolling through. No, a complete halt means no movement at all, with the nose of the car dipping down. Late one night I was woken by the horn of a car behind me. I must have stopped at a red stop sign while on autopilot, and I had stayed stopped there until the car horn woke me fully up again.

My 'stop at red' reflex became a tad too highly developed. Twice I remember pulling over to the side of the road to stop at a bright red letterbox — each time when I was tired and a fair way into a long shift.

Same rule for every single red light — I always stopped.

I would always put on my blinkers to indicate a lane change, even when I knew for sure that there was nobody else on a six-lane divided highway at 2 am.

I always put on a seatbelt, even just to move the taxi a few metres.

So I gradually conditioned my mind and body into these good habits and so wired an automatic pilot into my brain.

One night on my way home after a long sixteen-hour Sunday shift, I was woken by the blaring of car horns and the glare of headlights, which very scarily were on my left, and not my right! The road had curved left, but I had not. I had cut across the oncoming cars and was driving on the wrong side of the road, about to run into the kerb.

I was just able to recover without hitting anything and somehow got home safely, but with a rapidly beating heart.

Years later I was involved in a state government microsleep campaign. It was something I felt very passionate about, with my history of personal close calls while driving tired.

Bank business

It took a year of taxi driving to build up some street smarts. I worked out that on quiet nights in winter, when there weren't many people on the streets, the taxi radio was the key to getting fares. Back then, the deal was that the radio would announce a job, and the taxi with the closest location would get that fare. I knew that some of the big banks in the city had accounts with the company I drove for, and the staff would leave work in regular waves at around 8 or 9 pm. So I made a point of parking my taxi outside a bank around their finish time, so that I could be physically closest and successfully call on the job.

After a few weeks of regularly picking up the same bank employee, I got quite friendly with him. I'll call him 'Fred'. Back then there were no mobile phones for people to gaze lovingly at, so we always chatted for the duration of the long drive from the city to his home in leafy Pymble. We found we had an unusual (back then) shared interest in programming computers, and it turned out that Fred was a computer programmer in the finance department. We talked computers, which led us into artificial intelligence, science fiction, aliens and everything else. I looked forward to our slightly random and rambling chats.

After a few months, Fred simply vanished! When I picked up other bank employees, I would ask after my mate Fred, but strangely, the mere mention of his name was enough to make the other employees totally clam up and refuse to engage in any conversation at all. The less they said, the more curious I became, questioning every single ride until eventually, proving the old adage that hard work (sometimes) pays off, I struck paydirt. After asking my usual question, one bank employee responded with, 'F**king Fred! That bastard ripped us off for a million dollars.' A million dollars was a huge amount of money back then!

Apparently, Fred had realised that applying percentages to numbers could lead to tiny rounding errors. Each rounding error was very small (just fractions of a cent), but after a while, a huge number of tiny amounts could add up to a lot of money. Fred supposedly set up a program to collect all of those individual fractions of a cent and deposited them into a secret account. I guess that as the money accumulated, so did the risk of being busted.

Incidentally, Fred and I had a few discussions about the mathematical field called operations research or minimax, which involves maximising or minimising outcomes such as running costs, delivery times, kilometres travelled and so on. And we definitely touched on risk-benefit ratios, which was possibly relevant to how long he should keep siphoning money away. Maybe he was unwittingly using me as a bouncing board for his Big Heist? Or maybe, like me, he just liked minimax.

Regardless, once he had 'enough' in his secret account, according to his ex-colleague, he waited for a long weekend with the Monday off. He chucked a sickie on the Thursday morning and told his boss he would almost certainly not be able to come to work on the Friday either. Then he duplicitously boarded an international flight and flew to Switzerland, where he had sent his stolen money. He didn't turn up on the Tuesday after the public holiday either, so it was almost a week before his boss rang him at home to see how he was. Well, he was totally fine, but he was not at home to take the call. Fred had turned up in person to his Swiss bank on Friday Australian time, taken his money and vanished.

I never saw anything about it in the news, even though I looked. It was likely in the bank's interest to keep the details out of the public domain to avoid any copycat crimes – and just plain embarrassment.

> **White-hat hackers**
>
> A few years ago, when I was a speaker at a computer security conference, I went along to a lecture by Kevin Mitnick, an infamous early hacker. The movie *Takedown* is based on his story.

He got into computers in 1979 at the age of sixteen (the same age as I did) and then slid into hacking. He was eventually captured by the FBI in 1995 and served five years in prison for his crimes. On release, Mitnick became a computer security consultant for various government and private bodies, turning from a black-hat hacker (a baddie) to a white hat (a goodie). Typically, a white-hat hacker is paid to probe the security of a system – not for fraud, but to alert the company of any potential weakness.

After his lecture, he gave me his business 'card'. It was made from metal and had a set of picks to get into locks. My daughter Alice worked out how to use them and busted into a cupboard that we had lost the key for!

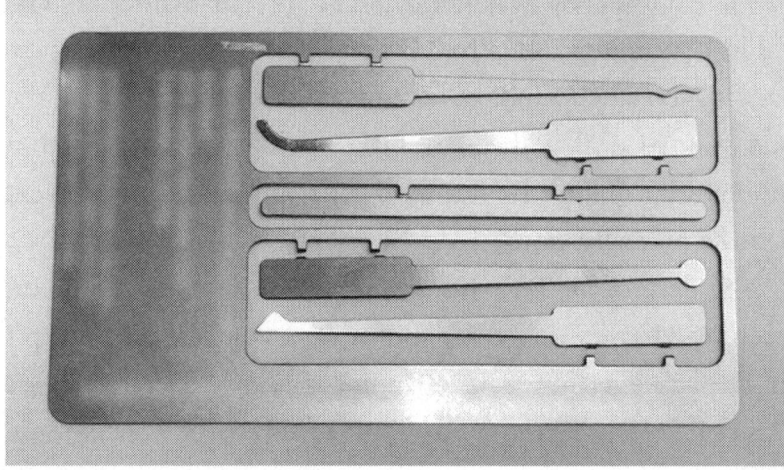

Million-dollar briefcases

For that little window of taxi-driving time, I skirted around the edges of the dark side. On my part it was voyeurism, along with intrigue. I didn't belong there, and I was lucky to see 'things' without being undone by them.

I had one hippie mate who fascinated me. He came from a wealthy family and had good looks, an amazing brain and a love of science. He made his money from random highly paid electronic studio work, growing some dope every now and then, and a seemingly endless supply of wealthy maiden aunts, who died every decade or so leaving him another tidy inheritance.

The first time I met him he wore dark red velvet pants with a purple velvet coat over a pure white ruffled pirate-style shirt that was cropped to his belly button. We were roughly the same age, but he was exotic! I was in awe of him. We came from such different backgrounds. He had family connections, property in the country and links all over the big end of town. In high school, he was going to his private school by taxi while I rode my push bike. He once shattered the cast-iron bath in his rented rundown country home, after deciding the best way to deal with a big brown snake in it was with a shotgun! I wouldn't be surprised if he still got his bond back, because he really was terribly charming.

Yet despite all our differences, we had many common interests. He introduced me to lots of new ideas and concepts that I had never heard of, including million-dollar briefcases. Once I tagged along with him to his mate's penthouse apartment in a very ritzy part of Sydney. I immediately realised, once we were in the private lift, that I was out of my depth by a long way. Once inside the lavish apartment, I saw wealth and luxury on a scale I'd never seen before.

Through the partly open door of one room we passed, I noticed about a dozen leather briefcases all stacked neatly under a table. Confused, I nudged my mate while silently pointing towards the stack. He unobtrusively put his index finger to his lips. He was implying that I needed to be careful.

In the entertainment area, we all got rapping. (Back then, 'rapping' had nothing to do with hip-hop but referred to deep and meaningful conversation.) I diplomatically asked our host questions about his line of work. He was happy to tell me that he ran brothels, but that the money from owning brothels was only okay, not brilliant – because he had to pay off both the crims and the cops. From his perspective, the main advantage to owning brothels was not the money but the connections. Over the years he had built up 'leverage' with his more high-profile clientele – it was all about power and protection, not just money. And he absolutely needed that leverage to keep out of jail because (as I later found out) he was also making a motza as a big-time drug dealer!

Once we were safely back in the car, I asked my mate about the briefcases I'd seen in the apartment. He reckoned they each contained a million dollars! A million? Sure, in new hundred-dollar notes, all from illegal drug-trading.

'What happens if he ever gets pulled over by the cops and they search the boot?' I asked.

My friend shrugged. 'He just rings his friend who's very high up in the police and the problem goes away.' There was a window back in the 1970s when the New South Wales police were said to have the dubious honour of being the most corrupt force in Australia.

It probably wasn't quite that simple, but it seemed shocking to me that he could apparently avoid getting busted with just one phone call!

The fuzz

But for us small-time hippies, getting busted was a genuine worry.

Early on as a taxi driver, the other hippie drivers had warned me about an undercover narc (American slang for a narcotics officer, from the drug squad). He dressed like a hippie (long hair, beard, flared trousers and tie-dyed clothes), talked cool, then after a while would try to sell you some dope – but if you bought any, he would immediately bust you for possession.

One day, this very same undercover officer got into my cab and tried to sell me some dope. I kept saying no – over and over. 'I'm driving. Look, man, I'm just not into it, cool?' He tried for a little longer but then gave up. When he got out, he immediately hailed another taxi. The word back at the base was that he spent the whole weekend bouncing from taxi to taxi trying to sell some of that dope (supposedly from a big bust in Wollongong the weekend before) to night-shift drivers. None of the taxi drivers in my circle fell for his act.

Illegal casinos

Around that time, illegal gambling was rife in Sydney. It was run very professionally and had no competition, because legal casinos did not exist back then in New South Wales.

The illegal casinos ran a roaring trade from the late 1960s to the late 1980s. I knew where they all were, because I drove passengers there and back! At their peak, the gambling dens raked in a combined total of about $600 million each year, which in today's terms would be about $5 billion. They were on a par with a big mining company in terms of income. (And like the mining companies, the amount of tax they paid was probably close to zero.) The Double Bay Bridge Club sat at the top of a flight of unobtrusive stairs on New South Head Road and took in the equivalent of $1 billion per year! (Today that same space is occupied by a dental surgery and a gym. I'm very confident that they don't have a revenue of $1 billion each year.)

On the one hand, the commissioner of police and the premier of New South Wales seemed terribly shocked, at least in public, at the suggestion there was illegal gambling anywhere in Sydney. And yet, every weekend, usually around midnight, people would jump into my taxi so I could take them to one of a dozen or so illegal casinos around inner Sydney, like the 33 Club on Oxford Street (handy to Darlinghurst police station), the Forbes Club in Woolloomooloo, the Goulburn Club in Haymarket, the Kellett Club in the Cross, and more.

Most casinos blended in with the surrounding streetscape on the outside, but on the inside they were luxurious, reminiscent of a scene in a James Bond movie. Knowing I was just a harmless taxi driver, the bouncers let me in a few times for a quick look around: plush red carpet, free alcohol and food, big fish tanks (they all had them for some reason) and, of course, blackjack tables, roulette wheels and a baccarat table or two. Everybody was well behaved and well dressed (apart from tourists like me). The clientele included local media and sport celebrities, doctors and lawyers, and well-heeled businesspeople.

For the uninitiated, one sure sign of an illegal casino was the line of taxis parked outside on weekends after midnight! There was never an official rank, but the taxis would accumulate because they knew the passengers would be coming. Sometimes I'd get a big tip from a successful punter. The cops never moved the taxis on.

One design element I really admired were the doors. At the 134 Club in Balmain, their very solid wooden front doors had a small sliding panel behind steel bars at face level, just like in the movies! Years later, when I was living in Lilyfield with my mother, I found several of these doors at the local timber yard, so I bought one. I thought it would be cute to answer a knock at the door by sliding up the 'casino panel'. To complement the casino door, I also installed functioning traffic lights at the bathroom door – 'red' obviously meant in use!

Doping horses

It wasn't just casinos that were big – so was taking a punt on the horses.

Late one Sunday night, I picked up a fare from the international terminal at Sydney airport. The guy who sat in the front looked about fifty, and we started chatting. He turned out to be what you might call a 'colourful racing identity' – larger than life, very wealthy and potentially a little 'bent'. Being simultaneously very naïve and very curious, I immediately asked, 'Is horse racing crooked?'

He laughed out loud and said, 'Pull over and I'll show you.'

'This is Alison Road,' I protested. 'There's no legal place to stop.'

'Don't worry about it. It's close to midnight, and anyway, I'm good mates with the cops.'

I put on the hazard lights and we got out of the taxi, and he pulled his briefcase from the boot and opened it up. 'What do you see?' he asked.

I had a look and said, 'A couple of hundred little glass bottles.'

He nodded. 'Right now, in the state of New South Wales, there's about a dozen drugs that you're not allowed to use on horses to make them go faster. But I've just come back from New York where there's about *two hundred* drugs that you're not allowed to use on horses to make them go faster. Right here I've got the ones left over, a hundred and eighty-eight drugs that aren't illegal – at least, not yet – in New South Wales. And you ask if horse racing is crooked? Ha!'

That guy had such a belly laugh at my expense. True story!

This next story is third- or fourth-hand, so I don't know how true it is. But it's a good one!

The great racehorse Phar Lap did very well in Australia before being taken to America, where he died under rather mysterious circumstances. His life was made into a movie, and of course Phar Lap's amazing 1930 Melbourne Cup victory was part of the film. The Flemington racecourse had been hired to recreate the scene, and so had hundreds of extras and their 1930s period clothes.

But according to the story that I heard, on the day of the shoot the director woke suddenly at two in the morning with a terrible thought running through his mind: how could they possibly arrange that the horse doubling for Phar Lap would win by a large margin, and that the rest of the pack would pass the post in the correct order trailing by the correct distances for the shot? I mean, these were animals, not trained actors!

In the morning, he called a meeting of the jockeys and asked. 'How can we make this happen?'

Baffled, the jockeys looked at each other before one of them said, 'What's the problem? We do this every Saturday.'

Anyhow, that's the story I heard ...

Smashing six cars in one go

When I was driving taxis full time, I was hungry for money, so I drove really, really hard and was constantly weaving through the lanes of traffic to get to the head of the queue. My meal breaks were short and sweet.

I had seen an advertisement on TV for a place called JumboBurgers selling hamburgers at an introductory fifty cents each – wow! You'd entice me, along with a lot of taxi drivers, with that kind of bargain pricing. The night was cold and foggy, and around 10.30 pm I pulled up on Canterbury Road and walked into the store, excited to bite into my

Above: Family and friends in 1949. My father, Ludwik, is centre front; I am the baby on his right; my mother, Rina, is at the back.

Left: Me as a young kid, c.1957.

Me in 1962 with disarmingly high pants and beloved camera plus anti-charisma effect. I can't see any other explanation for a female (even a family friend) being repelled away from me at such an angle.

Above: With my parents in 1966, at my graduation for my first degree, in physics and maths.

Below: An early bearded hipster, 1968. That drink with bubbles is not beer or coffee but cow's milk (which was the only milk option back then).

Right: Part-time Jesus figure, 1971.

Below: En route to an unsuccessful attempt to climb Mount Bangeta, New Guinea, 1971.

Bottom: Me in the highlands of New Guinea with my beloved Bolex film camera, wearing the infamous string shirt that got me my Nudge Nipples nickname, 1971.

Above: Yup, a literal flower-munching hippie, 1971.

Left: Channelling Tarzan staring into the distance, 1971.

Below: My taxi driver ID, 1972.

Above: Deviants do have more fun. Me in 1975.

Right: My 1976 hospitals ID. I worked at the Prince Henry Hospital as a scientific officer.

Below: On holidays in Ouvéa, New Caledonia, in 1977 with my parents, around the time my life path changed again and I became a biomedical engineer.

Above: I was roadie for The Magnetics (I'm wearing the T-shirt) in the late 1970s.

Below: Half a beard for half a while in 1981 – too scary to be out in public looking this odd.

Bottom: The Mitsubishi Lancer I had for twenty-one years, with my favourite waterbag, c.1985.

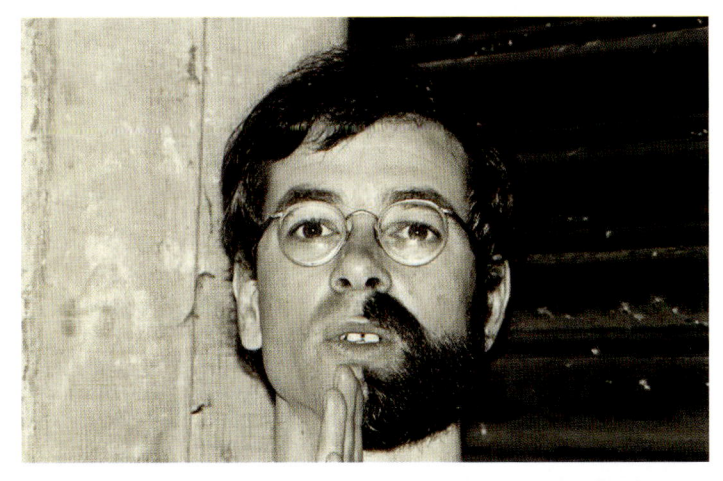

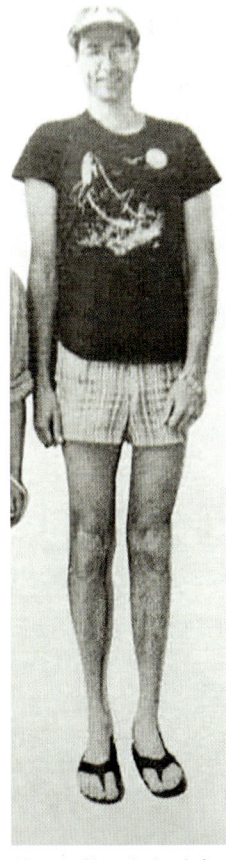

Above: Classic look for 1981 – me in short shorts and T-shirt.

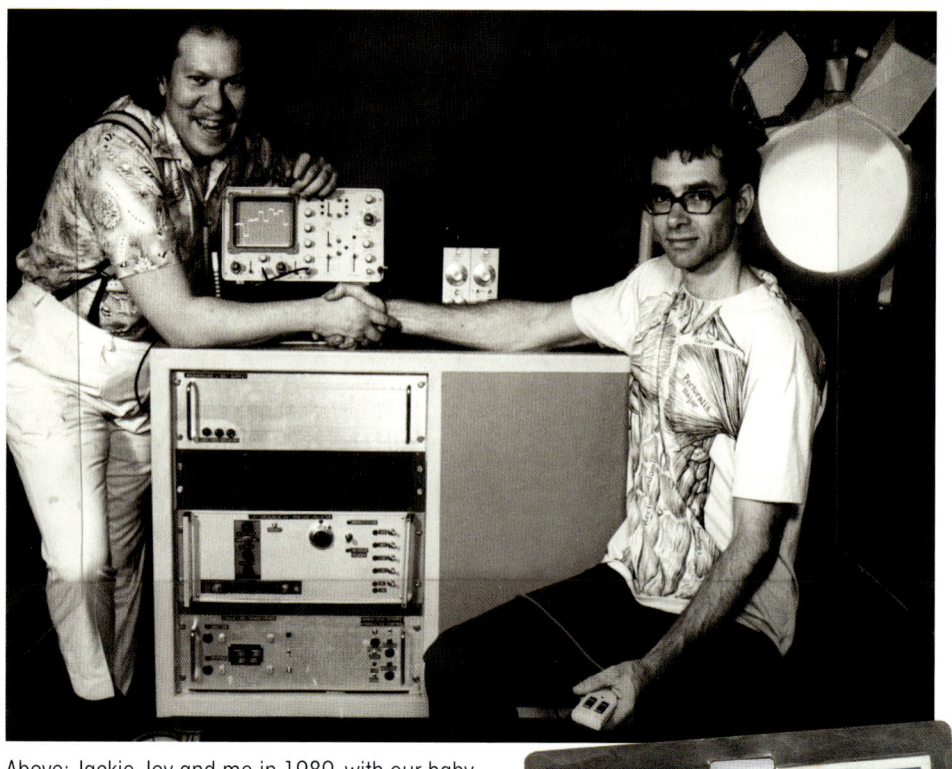

Above: Jackie Joy and me in 1980, with our baby, the electroretinograph (ERG) machine, after one year of thinking and nine months of planning and building.

Right: My kids' hospital ID.

Below: Me as the operator, with ERG machine and patient in comfy chair.

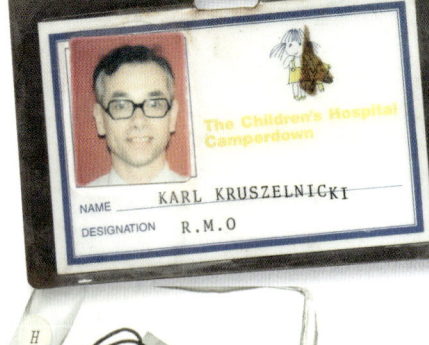

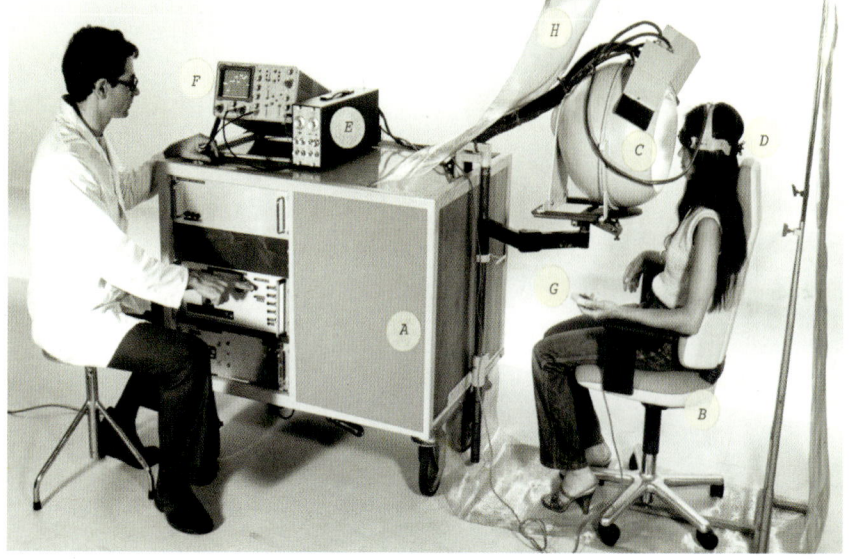

Above: Dressed up for Mary's Medical Graduation Ball, 1985.

Below: Mary and me, mid-1980s.

Above: In Fiji in 1987 at the same time as a coup was declared. Nothing was open and our last $2 had just been spent on beer by Mary!

Right: I was so lucky to be with Mary, c.1987.

Below left and right: Big Karl and Little Karl.

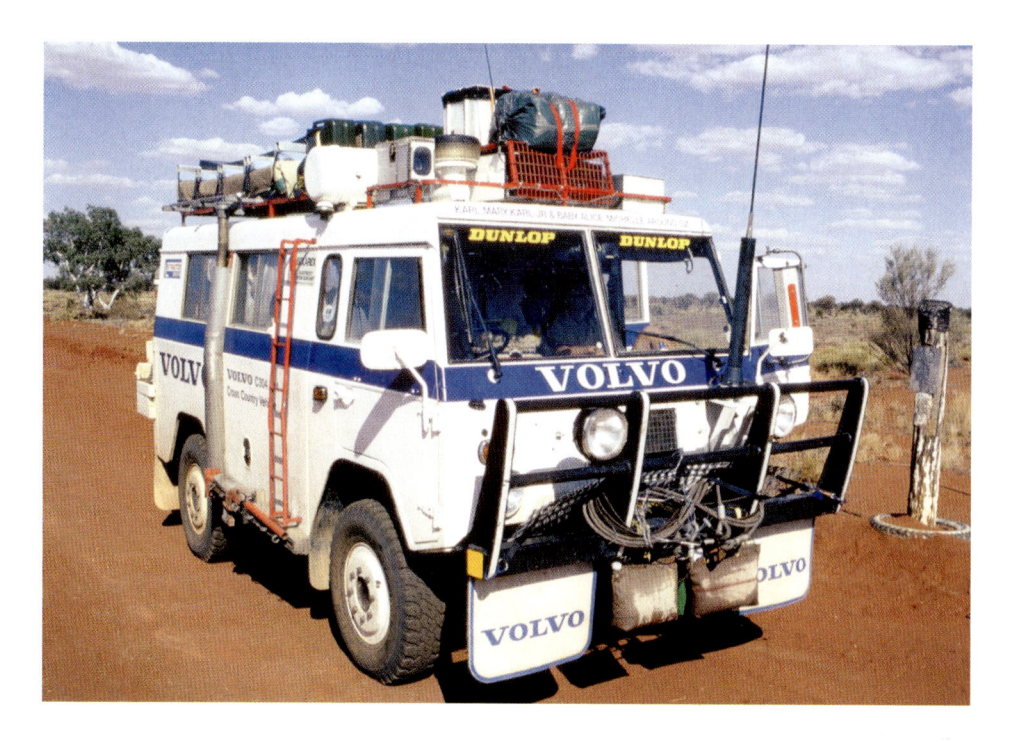

Above: On the border between the Northern Territory and Western Australia in our trusty self-modified Volvo C304 in 1990, on the lovely red dirt of the Australian outback (it's red because it's rust).

Below: Me as a 'bushman' in 1991, eating honey grevillea and loving the sugar hit.

Above: Sitting in the front seat of the fastest piloted jet ever made, the lovely SR-71, in 1996.

Below: My ABC IDs, with Lola, and Little Karl and Alice.

Above: Cuddling the official mascot of the ABC, a banana in pyjamas, in 2002.

Below: On Macquarie Island in 2010, on my first trip to Antarctica, surrounded by penguins, to whom I am reading excerpts from my 27th book, *Science Is Golden*. They are as astonished as I am.

Left: Alice and me, matching – like father, like daughter.

Below: In a mud bath with Alice in California.

Bottom left: Big Karl and Little Karl, both bald.

Bottom right: Lola and me.

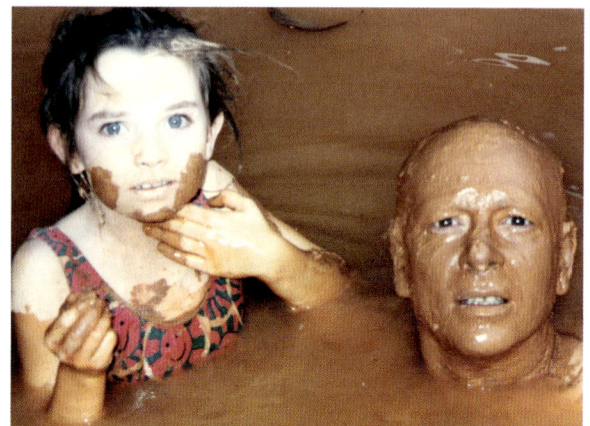

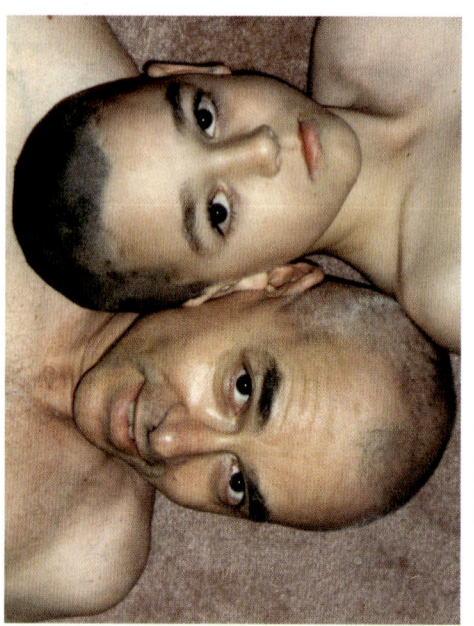

Above: Mary and me at our wedding in Norway in 2006, with Alice and Lola.

Below: Lola and me at the Great Barrier Reef, 2016.

Above: In the ABC Brisbane studio in 2018. After a third of a century of the pressure of live science talkback radio, I seem to have lost microphone fright. *(Isabelle Benton)*

Above: My birthday party in 2016, with my lovely children. The T-shirts referenced a famous Karl Lagerfeld meme – and the kids obviously thought I shared his great sense of style!

Right: In the main quadrangle of the University of Sydney, 2019. *(Ross Coffey, from* Who Do You Think You Are? Australia *© 2024 Warner Bros. International Television Production Limited. All rights reserved.)*

first ever JumboBurger. I got my burger and turned around to check on my taxi, which naturally I had parked in a no-standing zone. But my taxi was totally gone – no taxi! How could it just disappear into thin air? I checked my pockets to see if I had the keys – yep. As a taxi driver, it's considered very poor form to lose your taxi, and my heart sank. My next thought was that somebody had stolen it. Maybe I forgot to lock the cab?

With my JumboBurger in one hand getting colder, I stepped out into the night. I couldn't even take a bite because I felt sick to my stomach. Then who should I see but two detectives sitting in an unmarked car in the parking lot, JumboBurgers in hand! I rushed up to their window and said, 'Guys, somebody stole my taxi! Can you report it?'

They looked at me appraisingly, and then one of them said, 'What are you telling us for?'

Stating the obvious, I replied, 'Because you're cops.'

'What makes you think we're cops?' he shot back.

I stared at these two big guys, both wearing suits and sitting in a blue Falcon with a big V8 engine. Throw in the police radio sitting between them and they couldn't be anything *but* detectives, even without the JumboBurgers! Before I could think of a clever reply, they said, 'You're going to increase our paperwork.'

'Please,' I begged, 'just report on the radio that somebody stole my taxi.'

At that exact moment, a guy came running up the hill through the thick fog, shouting, 'Help! There's been a terrible crash. A driverless taxi just ran into a whole bunch of cars!'

Even before he finished speaking, the detectives had started up their big V8 and accelerated off into the night in a haze of rubber tyre smoke. If they wouldn't report one measly 'missing' taxi, then there was no way they were going to have anything to do with a multi-car smash-up! It would take hours to write up *that* paperwork.

I forced myself to walk apprehensively down the hill through the fog, and take a look for myself. Yes, there were half a dozen smashed-up cars, with my taxi at the head of the bunch!

My story (and I'm sticking to it) is that my taxi 'jumped the handbrake', meaning the handbrake mechanism had spontaneously released under spring tension. In the old days, cars had an actual handbrake lever that you pulled up with your hand, against the tension of the spring. Little metal pyramids in the handbrake mechanism created the locking mechanism, and as you pulled the handbrake on you could hear the little pyramids going clunk, clunk, clunk. With repeated use, the pointy bits of the pyramids would wear out, and the locking mechanism would let go. That's how you get your classic jumping-the-handbrake scenario. (I was taught to always push in the button on the end of the handbrake lever while I pulled it up, so as to not wear out the little pyramids.)

Anyway, what seemed to have happened is that my empty taxi jumped the handbrake, started rolling and then, having some momentum, it climbed the kerb. First. it knocked over the no-standing sign. Then it swung back onto the road and rolled downhill into the fog.

My taxi scraped down the side of a moving car that was coming towards it. That was followed by a head-on collision, which was then followed by three rear-enders. That made six cars all smashed up, including my cab!

JumboBurger still clutched in hand, I stared at the terrible mess. Eventually the uniformed cops turned up and took the details, but they didn't believe me about the car jumping the handbrake. Eventually I said, 'Look, I left the car unlocked,' desperately suggesting that somebody else might have released the handbrake.

'Yeah, well, it's illegal to leave your car unlocked.' (The cops can always get you if they really want.)

We went through all of the paperwork slowly and painfully. One cop said, 'Geez, mate, lucky nobody got killed.' Which was exactly what I was thinking!

A tow truck took me and my smashed-up taxi back to the base. The miserable part was that on top of smashing up six cars, I hadn't made enough to cover my pay-in (the amount you had to earn to cover the hire of the cab – after that, everything you made was yours). So I had to cover the pay-in out of my own wallet.

By this time, the other drivers had heard about my big smash and were laughing about me setting a new record for the most cars crashed in one night! I couldn't even go home. I had to wait until three o'clock in the morning for my mates to finish their shifts, so we could all drive home together.

The next day I came into the base for my shift. Let me paint the picture. It was a big garage, maybe the size of two

Olympic swimming pools side by side. Leaning against the walls were a few dozen drivers like me, waiting around in the hope that Ron, the boss, would give them a taxi to drive that shift. It was a real Depression-style set-up, with no guarantee you'd get a taxi, or any money, that night.

As I came in, Ron shouted, 'Ah, here comes bloody Nicky.' He couldn't pronounce Kruszelnicki, so he called me either Nicky or Wheelbarrow (the name he gave to anyone with a long or unpronounceable name). 'Bloody Nicky smashed up six cars last night with my taxi. And he thinks we're gonna give him another taxi! He's got to be bloody joking.' The other drivers laughed loudly at my expense.

I stood my ground. 'Yeah, it wasn't my fault. It jumped the handbrake because you don't do the maintenance and it was a shitty car.' Ron asked again what I thought I was doing there. I said, 'I'm going to start filling out the paperwork and then you'll give me a taxi!'

Ron fired back, 'You're not getting any bloody taxi, but you do have to go and fill out the paperwork.'

I went upstairs to write out the forms (I still hate filling out forms) and then came back down. Ron looked up at me and said, 'You know you're not getting a taxi today. They're all gone.'

But I just kept coming back – Friday and then Saturday and then Sunday. And I kept on coming back week after week until finally Ron said, 'All right, take William – he's in the workshop.' I should have realised that there would be a twist with William ...

William the Conqueror

The taxi-base manager was remarkably well read. We both knew that William the Conqueror invaded Britain in 1066. William turned out to be a very beaten-up taxi, with, of course, the numberplate 1066 – which, going by how beaten-up he was, could hypothetically have been his year of manufacture.

In the workshop upstairs, I told the mechanic I was driving William. He dubiously told me William was 'mostly ready' then gave me the keys, shrugging his shoulders as if to say, 'It's your funeral.'

The engine sounded fine. So far so good. But as soon as I drove forward, it was clear that the steering was very wrong. William pulled really hard to the left. To drive straight ahead, I had to constantly pull the steering wheel to the right. Okay, I could handle that. But on the steeply curving down ramp to exit the base, as soon as I touched the brakes, William suddenly swerved violently to the right!

Because I really was a good driver, I got William down onto the street, despite the combined brake/steering problem. William was clearly complicated: normal driving had him pulling hard to the left, but touching the brakes had him pulling hard to the right. But that seemed manageable, if that was the extent of his issues. That turned out to be a big 'if'!

There was a cab rank down the bottom of the street and four guys were waiting there, looking panicky. I pulled in and asked, 'Where to?'

'The airport,' they said.

'Hop in.'

They tried the front passenger door. It didn't open! Then the rear passenger door didn't open. They had lots of luggage, but you guessed it, the boot didn't open either. The only door in the whole car that opened was my driver door! Luckily, the windows wound down, so they all climbed in through the windows and stacked their luggage on their laps. It must have looked ridiculous from the street – and they must have been really desperate for a ride. I somehow managed to get them to the airport, despite William's many idiosyncrasies – but I strongly suspect they will have remembered this trip to the airport their whole lives too!

I drove that entire night in a taxi that every single passenger had to enter and exit through a window. As far as cabs go, William pretty much conquered me! Ron begrudgingly admired my determination and eventually gave me better cabs.

Driving too hard

Gradually I worked my way back into Ron's good books. One night, I got an almost brand-new cab. Oh my God, it was beautiful. The Ford Falcon XY was a taxi with a very powerful engine. Back then there was little safety gear (such as anti-intrusion bars, safer brakes, airbags, etc.) so the XY was very light, really fast – and not very safe.

Taxi drivers loved a bit of 'grunt', and we got into the habit of seeing how fast we could go on the straight runs – which wasn't very wise at all. We could push the older taxis to just over 140 kilometres an hour at about two in the morning when there was no other traffic. But in this brand-new XY cab, blow me down, you could nearly get to 160 kilometres an hour.

I fanged the heck out of that beautiful new car for the entire twelve-hour shift. Even though it was an automatic, as per my normal, I was using both feet – the left foot for the brake and the right for the accelerator. The advantage of the two-foot technique was that it gave me more precise control, and I could quickly slip into any tiny gaps that opened up in a stream of cars.

When I came back the next day to get a taxi, Ron said, 'Hey, Nicky, you know that brand-new car you got last night? Well, it was nice and tight before, but the owner says the body's all creaky now.'

Wouldn't you think a car should be able to stand one night of hard driving?

Lost in the Yagoona tip

It was a bit after midnight, and I'd had a longish stretch waiting at the taxi stand. A passenger hopped in and asked, 'Can you take me to Yagoona?'

This was a good long ride and a decent fare, even if it meant returning empty to the city. 'Sure, no worries,' I said.

He said he knew the local roads close to his home really well and offered to show me a shortcut. 'We can cut through the Yagoona tip,' he said.

Right on – I loved a new shortcut! Sure enough, very soon we were driving along makeshift dirt roads between stacks of piled-up garbage, darting this way and that. He knew them like the back of his hand and gave me clear directions about where to head next.

Before driving through its gates, I had no idea how huge and random the tip would be. I was surprised that it didn't really smell bad, but maybe that was because the night was pretty cool. It was exciting exploring somewhere I'd never been before, even if it was a rubbish tip.

After dropping off my fare, I decided to return the same way seeing it was such a fantastic shortcut. As I tried to retrace my route through the tip in the pitch dark, I quickly realised I was going in circles, getting more and more lost. There were so many possible turns. The street directory was no help – in it the tip was just a huge beige area. It should have been signposted with something helpful like 'Here Be Dragons'!

No map, no probs. I'd call the taxi base on the two-way radio and get them to talk me out. Good idea but no dice; turned out that the Yagoona tip was a complete dead spot for reception. As things went from bad to worse – no map and no radio – I thought about navigating by the stars. But of course, it was a thickly clouded night without a single star in the sky. On top of everything else, I realised I was nearly out of petrol, which meant I couldn't keep randomly cruising around trying every possible combination to find a way out by chance.

I was genuinely stuck in the Yagoona tip!

My biggest worry was not making my 3 am changeover, when I was supposed to hand the taxi over to the next driver. Instead, it seemed I was going to run out of fuel or be buried alive by a massive pile of garbage toppling on top of me!

Out of escape ideas, I pulled over and turned off the engine, waiting for the first hint of daylight to guide me out. I passed the hours thinking about all sorts of things, including how that guy knew his way through the tip. Had he played here as a kid?

Eventually the first hints of dawn showed me a soft glow in the east, and I navigated my way out by heading towards the rising sun. I rolled back into the base about half past six, with everybody who was waiting on the car looking really annoyed: the car washers, the petrol pump attendants and especially the day driver were all seething.

I started my defence with, 'Look, I'm really sorry, but I got lost in the Yagoona tip.'

I was kind of expecting an outburst of sarcastic laughter, but oddly, they nodded sympathetically and said, 'Yeah, yeah, happens to the best of us.'

Beaten unconscious

One night I was about halfway back to the city after a good long run out west. I wasn't expecting a return fare, but I was always hopeful (irrationally optimistic, that's me). To increase the slim odds of hearing anyone shouting 'Taxi!', I was driving with the windows down (and there was no air con in those primitive days).

Out of the blue, I thought I heard someone call out, and I jammed on the brakes. At first I couldn't see anyone, but then, in a vacant lot nearby, I saw a group of people. A person was running towards me, while being chased. It turned out to be a woman, who jumped into the front seat of the cab in a complete panic and shouted, 'Drive, drive, drive – just go!'

I was shocked but hit the accelerator. I had barely got us moving when one of the men chasing her reached my taxi, grabbed her door handle with his left hand and swung it open.

Everything was happening very quickly. And everyone was shouting.

He wouldn't let go, and she was screaming at the top of her voice. Because I was accelerating, and he wouldn't let go of the door handle, the forward motion of my taxi swung him towards the car. He shoved his right hand on the pillar between the front and back passenger doors to stabilise himself, but then the door closed on his right hand and he started screaming in pain. For some reason, I couldn't stop thinking that I was going to tear his fingers off.

It was a terrible situation, and I really did not know what was going on. I stopped the car to try to work it out – and that turned out to be a big mistake.

I hadn't noticed that all his mates were very close behind us.

The moment the taxi braked to a screeching halt, they were there at my door. They ripped it open (I wasn't quick enough to lock it), dragged me out and started beating me savagely. There was no way I could fight them off – there were too many of them.

Very soon, the pain I was feeling over my whole body passed from a terribly localised into an almost ethereal pain, as if the pain had shifted like a halo to the outside of my body. As I descended into unconsciousness, I dissociatively wondered if they would stop beating me before they killed me. But it was all totally out of my hands. I had no control over anything.

After about ten minutes I came to, lying on the road. I was nursing broken ribs, but luckily, nothing worse than that. The taxi was idling next to me, and my first stupid thought was, 'At least they haven't stolen my cab.' My next thoughts were 'What happened to that poor woman?' and 'Thank god they didn't kill me.'

I should have gone to hospital, but at that stage of my life I had no medical training or understanding of how dangerous being violently knocked unconscious could be. Instead, I drove to the nearest cop shop to report what happened.

Later the cops told me that the woman survived, but bad things had happened to her. The police never found the guys who did it. I was anguished that I didn't manage to get that vulnerable woman away from those violent men.

After that, I was a bit more wary. It didn't stop me from pulling over the next time someone called out for a ride in the dark. But I was just more ready to lock my door if something bad seemed like it might happen.

Nathrurd

I loved driving on Sunday mornings. It was a nice long shift, usually from eleven in the morning until 3 am on Monday. So I had lots of time, and the day wasn't too rushed. I knew I would definitely make my pay-in nice and early, so I could relax.

Generally I gave a wide berth to the characters who'd been out all Saturday night and were now heading home, still drunk and often a bit aggro. Instead, I'd deliberately take my taxi to the suburbs where families, all dressed up in their best, wanted a lift to or from church, or to visit friends and relatives. They were so polite and friendly. It was such a good start to a shift.

One Sunday I picked up a smiling passenger who, when I asked where he wanted to go, said something like 'Nathrurd'. I'd never heard of the street, so I told him, 'Sorry, mate, I don't know where that is.'

I quickly realised he spoke no English at all, but he was prepared for this. He pulled a bit of paper from his pocket where he'd carefully written the name of his street: *No Through Road*. I had to laugh.

It floored me for a minute, but I really wanted to get him home. I worked out that he spoke Italian, so I drove to a nearby church that I knew was patronised by a lot of Italian-speaking Australians. I got one of the churchgoers to act as an interpreter. Together, we asked him to draw a little mud map of his street and the surrounding streets, which helped narrow the location down to just two suburbs. So I set off to take him to the half a dozen or so possible locations of his home, one at a time.

He was absolutely thrilled when, on about the third try, we found *his* street. Before he got out of the cab, I wrote down the name of his street and suburb on that little bit of paper he was carrying.

He wanted to pay me for the trip, but it was such a happy ending that I was satisfied with successfully solving the problem and finding his house. He was a stranger in a strange land, and I wanted things to go well for him.

I drove the rest of the fourteen-hour shift on a completely natural high!

To the Gap and back

One busy Thursday arvo, just before sunset, I picked up a passenger who asked me to take him to the Gap, about 30 kilometres to the east.

The Gap is a cliff in Watsons Bay right on the ocean. Besides being a tourist attraction with great views in all directions, it is also a suicide destination.

A local, Don Ritchie, kept an eye out on the edge of the cliff for decades (until he died in 2012). He was nicknamed the Angel of the Gap. Since 1964, he had talked several hundred people out of their suicide attempts, by simply caring enough to engage them in conversation and invite them back to his home for a cuppa.

My passenger was smoking in a curious way (yes, back then, you could smoke in a taxi). He seemed simultaneously both determined and relaxed, as though he had finally made a difficult decision. And I was worried that decision might be suicide.

When we arrived, I switched off the engine and immediately offered him another cigarette. I also turned off the taxi meter, obviously resetting it to zero. He said yes to the smoke, so I rolled him one and we sat in the cab for ages just talking and smoking. He seemed to settle deep into the seat, which I was very glad to see, especially if the alternative was him getting out and heading straight to the edge of the clifftop.

After more than an hour, he said, 'Look, you know what, can you just take me home?'

'Sure,' I replied straightaway. And then I asked, 'Were you going to jump?'

'Yes,' he said, 'I was going to jump. But it was just in the heat of the moment.'

I drove him all the way back and dropped him off near where I'd picked him up. I hadn't turned the meter back on, and when he asked me how much the fare was, I said, 'Don't worry about it.'

I hope my passenger eventually felt better, and was also able to share a laugh with someone close to him.

Wheelman

Early one Saturday night, I picked up five pretty big and fairly rough-looking guys. (Taxis then had bench seats in the front, so they could carry three passengers in the back and two in the front.) I took them from the Inner West into Kings Cross. We were close to their destination, in a skinny one-way street with cars parked on both sides, when a car pulled straight out from the kerb without indicating and rammed straight into the car directly in front of me.

I called out, 'Hang on,' and braked hard, screeching to a halt. I realised this was going to be drawn out and messy, given the smashed cars blocking the road, but there was no other car behind me so it was time to hit reverse!

Using the official High-Speed Reversing Method as taught to me by experienced taxi drivers, I first flicked on the hazard lights. Pivoting my body towards the front-seat passengers, I slung my left arm across the back of the bench seat and turned to look out the back window. Time was of the essence.

The great thing about the High-Speed Reversing Method is that the back end of the car will go in whatever direction you move your right hand on the wheel. Move your hand to the right, and the back end of the car goes to the right. Move it to the left, and the back end goes to the left. No thinking required.

With one foot on the accelerator and one foot on the brake, I drove backwards at high speed for about 100 metres, concentrating hard on not running into the parked cars either side of me. I really wanted to get out of this one-way street before another car came along and blocked me in.

In just a few seconds I was back on the main street and was relaxed enough to talk to my five burly passengers, who were looking rather impressed. 'I'll have to go another way. You cool with that?'

They said sure, and as I dropped them at a sleazy illegal strip joint/casino, one turned back and said, 'We need a wheelman next Tuesday night. You interested?'

I knew enough criminal slang to know the wheelman drives the getaway car. And I knew I was definitely not interested, for at least three reasons.

First, I really didn't want to get involved with criminal heavies. I had absolutely no interest in jail time.

Second, they were asking me, a random taxi driver they'd never met before, to get involved with their heist. I figured it wasn't that smart to share the details with any passing taxi driver, and if they weren't that smart then maybe they would get busted, taking me with them. Again, no interest in jail time.

Third, I had long ago given up driving on Tuesday nights. It was the worst night for fares, and I didn't want to start driving Tuesday nights again for anyone, not even a bunch of heavies.

But there were five of them and one of me, so I let them tell me all the details about where and when to meet them to drive them to the 'burg' (that's crim talk for 'burglary').

I took the easy way out and chose the no-show! I really didn't think that these crims would ever bother tracking me down. I knew nothing about them, and they knew nothing else about me.

Taxi driving was a rite of passage for me. It was formative, allowing me to expand my view of the grown-up world, and then make a better version of me.

In certain ways it was very risky. One driver I knew well was killed for a lousy forty dollars, and another was beaten to death in the middle of the Sydney CBD. But there was a sense of camaraderie with other drivers, who were non-judgemental of my hippie lifestyle.

It turned out to be a good way for me to earn my living for several years.

1971.5–75

Disturb the universe

Up until now, I had focused on developing my *brain*, not my *hands*. By this time I'd finished my university education and had worked as a physicist. I knew some big-picture stuff, such as how the universe began, but I couldn't even adjust, much less fix, the handbrake in a car. I was much too 'theoretical'.

There are two types of physicists. Theoretical physicists are very good at abstract ideas, but usually have no clue how to build the machines to test them. Experimental physicists have the skills needed to build these machines but don't usually have the big revolutionary ideas. The theoreticians and experimentalists need each other, but the trouble with the former is that they are so abstracted from the practical world that they carry a 'confusion field' with them wherever they go. This means they can't interact with physical things, and sometimes the confusion field goes further and accidentally destroys things just by being near them. So the joke is that whenever theoreticians walk into the photocopy room, all the photocopier machines die on the spot.

On the spectrum between theoretical and experimental, I was closer to the useless end. But over the years I began picking up various practical skills: mechanical, electrical, electronic and the like. And it was my love of cars that started the transition for me into developing physical skills.

After I left New Guinea, I arrived in Sydney with no car. I knew that hippies drove VW Kombi vans, so I went to my Wollongong VW mechanic, keen to solidly build my hippie image even further. Amazingly, he had a Kombi right there in the shop that I could buy on the spot. I wrote him a cheque and I started driving back to Sydney, which started with a steep climb up Bulli Pass.

That bomb of a car stalled when I hit the steepest stretch of the pass. I couldn't understand it. I rolled backwards down the hill to a flat section then tried again with a bit more speed, over and over. But no, every time I got to the steep bit, the Kombi would inexorably slow down – all the way to zero! Of course that meant I was rolling backwards while dodging cars going up the hill, with engines that did work! It didn't make sense to me. How could a car not be able to drive up a hill?

I could get to Sydney via Mount Ousley, which was not quite as steep. The Kombi just managed that but at the less than blistering speed of 5 kilometres per hour, which was about as fast as walking! Even with my limited mechanical skills, I diagnosed there was possibly something very wrong with the engine.

So I drove back and told my mechanic the problem. I thought he was an honest person, and I naïvely didn't realise that he knowingly sold me a car with a totally clapped-out engine.

He was very happy to change over the engine on the spot, free of charge! He just happened to have a bunch of other clapped-out engines lying around, and it was an easy fix for him. With the early Kombis, it was ridiculously quick to replace the rear-mounted engine. You could do the swap in fifteen minutes, without rushing. To remove, undo the bolts holding on the back bumper and lift it away, support the weight of the engine on a mobile jack, undo the four mounting bolts holding the engine onto the gearbox, crack apart the fuel and electrical connections, and pull on the jack to roll the engine back. To install, just reverse the above. So in a quarter of an hour, he put in an ever so slightly less gutless engine and sent me on my way. This one could make it up Bulli Pass – with a top speed of 15 kilometres per hour. So better than before, but not much.

I didn't realise it back then, but neither a Volkswagen Beetle nor a Kombi count as Real Cars. When you ride a bicycle powered by your legs, you have to slow down when you're pushing into a headwind or going up a hill. You accept that on a bike. But if your car powered by a motor slows down when you head into the wind or drive up a hill, it's not a Real Car.

On the other hand, the Kombi vibe was ultra-cool in the surfie and hippie worlds, probably because you could take a lot of your friends with you. But on the other *other* hand, the Kombi was horrendously underpowered, so you couldn't accelerate out of trouble. Going up hills at my top speed, I nearly always built up a queue of impatient drivers behind me who quite reasonably wanted to drive around the speed limit, not 30 to 50 k's below it.

The Kombi was also ridiculously and fundamentally unsafe. (Read Ralph Nader's 1965 book *Unsafe at Any Speed: The designed-in dangers of the American automobile* to learn how bad cars used to be.) The suspension on the back wheels was the cheap but inherently unsafe swing-arm suspension, which was fine if you drove only in straight lines. But major problems arose when you drove a VW Beetle or Kombi moderately quickly around a corner. In fact, really bad stuff happened – like tipping over!

My final showdown with the Kombi came a few years later in a slow-motion collision.

I was in Sydney traffic, slowing down behind another car. Fortunately we were both rolling at less than a walking pace. At one moment, my right foot was pressing comfortably on the brake pedal to ease me to a halt. In the next moment, something inside the hydraulic brake system haemorrhaged. The brake pedal lost all resistance, and my right foot immediately and effortlessly went all the way flat to the floor! Hitting the brake pedal to slow me down was as effective as whistling out the window. Less than a second later, at very low speed, I crashed into the car in front of me.

I was going so slowly that there was no damage at all to the car I hit, but the Kombi was a mess. In a Kombi, there's no engine in front of the driver. You sit right at the front with a stupidly thin sheet of metal between you and the outside world. The distance between your toes and the exterior of the Kombi is maybe 20 or 30 centimetres.

Worse, the front of my Kombi was no longer made from the metal it was produced with in Germany. That original metal had at some point rusted out, and my shonky Wollongong mechanic went for the quick fix. He had filled the gaping hole with plastic flyscreen mesh onto which he had slapped some 'bog', a plastic-based filler with the structural integrity of cardboard. As we collided, a metre-wide section of the front of my Kombi disintegrated in a shower of rust and plastic powder that covered my body. The front end collapsed like tissue paper and my toes ended up just a few centimetres from the back of the car I ran into!

If I'd been travelling at even 10 or 15 kilometres per hour, I could have lost my legs.

Chevy made me a mechanic

By coincidence, at that time a friend of mine was selling his 1954 Chevrolet and offered it to me for twenty dollars. It needed a fair bit of work, but at least it didn't slow down in a headwind or going up hills. So far so good!

That car changed my life. By learning how to fix the Chevy, I became a pretty capable backyard mechanic.

But first I needed to read the workshop manual, not easy to get for a twenty-year-old car. I looked fruitlessly in the spare parts departments of car dealers and at car accessory shops. Eventually I borrowed one from the Parramatta library and then did something very naughty – I stole it.

To get the library card in the first place, I needed ID. I didn't want to use my taxi licence as ID for various reasons (like unpaid road fines). I had already manufactured a fake identity for my home phone connection – Michael Anthony Conroy, or 'Mac' for short. Maybe I subconsciously devised an Anglo nickname to compensate for being a refugee and being bullied in school, I don't know. It was ridiculously easy to get fake IDs back then – so easy that a friend of mine got bank accounts set up for his dog Rinso!

So I registered at the library as Michael Anthony Conroy with a fake address, borrowed the book and then hung on to it for over a year, unavoidably making it dirty and greasy. I'm still embarrassed by the petty theft, but I eventually restored the balance.

One night around 2 am, driving my taxi back from Manly, I noticed a shop on Military Road selling car workshop manuals. And lo and behold, there in the window was a genuine workshop manual for a 1954 Chevrolet! I did the right thing and I went back in business hours and bought the manual. I mailed it to the Parramatta library, hanging on to the old grease-stained manual for myself.

With that manual, I started to learn how to fix up my new (old) car. It was so exciting – a journey to the mechanics equivalent of the Cave of Wonders.

First, the Chevy was mechanically very simple, and there was lots of room to work – unlike the Kombi, whose mechanicals were all squashed up and lots of special tools were required. But not the Chevy! There was so much room that when I needed to work on the engine, I would simply open the bonnet and sit on the side panel with my legs physically inside the engine compartment.

Second, the 1954 Chevrolet manual was The Best. It assumed I knew nothing and took me gently by the hand all the way. In the first few pages, for instance, it spelled out how to use and read the dipstick for the engine oil: 'The oil gauge (Fig 2) is marked "Full" and "Add Oil". The oil level should be maintained between the two lines, neither going above the "Full" line, nor under the "Add Oil" line.' It even had arrows pointing to the lines – how good!

It was like having my own personal experienced car mechanic taking me through each step, with infinite patience and knowledge. It told me that short trips would generate corrosive sulphuric acid, which would accumulate in the engine oil and 'eat' the internals of the engine, such as the main bearings. However, on a longer trip, the oil would get hotter, and the sulphuric acid would evaporate (presumably going into people's lungs), leaving the engine safe. So short trips meant the oil would have to be changed more frequently. Very useful!

Each section started off with an easy overview of what that bit of the car was supposed to do and then followed up with detailed explanations. It was even poetic! On page 0-7 (General Lubrication), for example, in warning of problems when grease is used in excess on a door, it explained that 'a slight brush across this soft grease may ruin a gown and spoil the entire evening for the owner and others'!

By comparison, in the Body Lubrication section, the Volkswagen workshop manual might say something like

'Lubricate the door hinges' – no information about what kind of lubricant, how much or how often.

Thanks to that Chevy manual, I got good enough to slowly work my way through the entire car, fixing everything to original specifications, which included front and rear suspensions, wheel bearings, brakes, engine and the drive train, steering, hanging the doors square, lubricating the door hinges and more.

As I built up my skills, I also built up my mechanical tool kit. I bought each tool as I needed it, which was not the most economical way to put a toolbox together. If I was unable to fix an issue in the morning due to a missing tool, I would buy that specific tool at midday and have the job completed by sunset. It gave me immense satisfaction.

I fell in love with the concept of tools as 'force multipliers': they could help you do things that you couldn't do with your bare hands alone. You could change something in the world around you from 'this' into 'that' – so you could 'disturb the universe'. I loved being free from helplessness and having to rely on others to repair things. (Being mechanically independent is different from being emotionally independent. I love having emotional connections to people around me.)

By the time I finished with the Chevy, I could drive it on a deserted dead-flat road at 145 k's per hour (its top speed), take my hands off the steering wheel and the car would track in a straight line for hundreds of metres. Furthermore, there were no vibrations anywhere in the car at that speed because I'd had the tyres and entire drive train properly balanced. And finally, when I braked from 145 kilometres without my hands on the wheel, the car would stop dead straight because I had adjusted the drum brakes perfectly.

But achieving all that was no overnight job. I drove taxis on weekends to earn money and spent the rest of my time fixing the Chevy. I did no filmmaking. Being a roadie? Sure, if the band needed me, I'd do it. But the rest of the time I was going through that car from front to back, making everything absolutely beautiful.

It took me several months to fix, but the skills I learned on the Chevy could be applied to any car (including a stupid Kombi). Some fifteen years later I needed those skills to modify a military vehicle (a recoilless rifle carrier) into a virtually unstoppable four-wheel drive vehicle (that led to my intermittent and quarter-century career as a test driver for four-wheel drives).

I even spent time as a backyard mechanic, fixing up cars for money, but I only did about half a dozen before giving up. It was intellectually satisfying to fix cars, but I could make a lot more money, and a lot faster, being a taxi driver.

And then, almost accidentally, I added electrics and electronics to my skill set.

Mechanics to electrics and electronics

I moved in with a mate who had a 1959 Chevy, working as an unpaid apprentice in lieu of paying rent. I was happy to work for nothing and learn from him as I had enough money to get by from driving taxis. I learned a lot of car mechanics from him as well as electrics and electronics. (Electrics is plain old AC and

DC power – while electronics uses that power to run circuits that can make simple decisions like turning something on or off.) He was doing amazing artistic stuff that was simultaneously creative and at the cutting edge of electronics. For example, he made wall-sized fluorescent perspex sculptures that were beautiful in themselves but also had electronics directly and visibly mounted into them, no boring circuit boards. Back in the early 1970s, these electronics were amazingly advanced, with integrated circuits and light-emitting diodes (LEDs). Strings of LEDs lit up in different ways, with the colour and brightness changing with the music and creating patterns that swept across the LEDs like bands of colour over the body of an octopus. Various wealthy art collectors and hip restaurants commissioned him to build them.

I enjoyed being an apprentice. It was easy: just follow instructions. He taught me the skills of delicate soldering, heavy soldering and basic electronics. After a few thousand solder joints, I got fairly good at it. I learned electrics from him as well. This was essential because his sculptures got their power from the 240-volt electric mains, not from low-voltage DC batteries that would go flat after a few days, so he needed to wire in a transformer to drop the potentially lethal 240-volt electric mains down to the safe low-voltage DC that the electronics ran on.

It is illegal for an unqualified person to touch anything related to the 240-volt mains electricity because it can kill you. And that's a very good reason! I was not formally trained as an electrician, but my friend was very safety conscious and taught me well.

240-volt electricity can kill

One very important lesson I learned was that when dealing with unknown electrics, always keep one hand in your pocket. What that meant was – do not touch an unknown electrical circuit with two hands. And in turn this means do not let 240 volts run through your heart, because that will interfere with the natural electrical signals that run your heart and maybe kill you.

Here's how that works. Your heart pushes blood through your body roughly once every second triggered by electrical signals.

First, a pulse (about 80 millilitres) of high-pressure oxygenated blood is propelled from your heart by its muscles contracting. The many different muscle fibres in your heart squeeze in an exquisitely synchronised series of contractions, to push that pulse of blood out into your general blood circulation at a pressure of about 120 mm Hg (which stands for millimetres of mercury, the measure of blood pressure).

That high-pressure blood forces various pressure sensors (baroreceptors) in your blood vessels to relax. But once the pulse of high-pressure blood passes by and your blood pressure drops to about 75 mm Hg, the baroreceptors fire. Then they emit electrical signals that are carried to the various muscles of the heart, which contract again. And so, about one second later, another pulse of high-pressure oxygenated blood is pushed into the general circulation of the body – and you live for another second.

So the 'normal' state of affairs is that a pulse of electricity arrives at your heart about once every second. This gives enough time for the heart muscles to contract and then relax, ready for the next electrical signal.

But if 240 volts of electricity travels between your left and right hands, it's also going through your heart. This 240 volts switches on and off fifty times each second, which is a lot faster than your heart's natural rate of once per second. The many thousands of individual muscle fibres in your heart cannot fire fifty times each second. So they either switch off totally, or go into an uncoordinated random firing called 'fibrillation'. According to heart surgeons, who at times during surgery will physically hold a fibrillating heart,

A Periodic Tale

> ```
> it feels like holding a 'bag of furiously
> wriggling worms'!
>   Either way, no blood leaves your heart,
> and Bad Things happen to you.
> ```

My Chevy mechanic electronics sculpture artist housemate was also a muso who had a degree in architecture. Further down the line I 'collaborated' with him in a competition to design the new Australian Parliament House. It was a two-stage competition, with the winners of the first round going to the second round. There was a lot of historical reading involved. It turned out that some parliament houses tended to follow the cardinal directions (north–south and east–west) for the main axes, perhaps related to crucifixion and Christianity metaphors.

But what my housemate came up with instead was something that I thought was rather clever: a circular and intimate design that allowed close contact between the various political parties. The winning design turned out to be exactly the opposite. In fact, some politicians have complained of the physical distance (and therefore lack of chance meetings) between the members of different parties, or even between the executive and lesser members of the same party.

I have no architectural training, so my total involvement was working out the numbers – basically, being a quantity surveyor. The building site was a hill, so my job, once I had the design, was to work out how much dirt had to be shifted. (By the way, the winning design involved shifting a million cubic metres of dirt and rock!)

We didn't win, so that project just died a natural death. And my life didn't go down the pathway of becoming the project manager for the construction of the new Australian Parliament House!

I moved from my mate's place soon after. My next home had very little resemblance to any parliament house, new or old. Towards the end of my hippie years I moved into a squat (!) in Glebe – more on that next. I met and was befriended by a Glebe local who, by the time he was sixteen, had stolen some two hundred Holden Monaros, a two-door coupe that usually had a very powerful V8 engine. It was the first time I had met anybody who had stolen even a single car – much less two hundred! But by the time I met him, this was all in his past.

I was still pretty naïve about the Real World, and there was a tough side to Glebe in the 1970s that, even as a taxi driver, I had been totally unaware of. I learned to recognise and give a wide berth to the very mean Mitchell Street Gang. My new bestie was actually pretty tough and could take care of himself, so there was an uneasy but respectful truce between him and the gang.

My reformed car thief came from a very dysfunctional family. But he managed to keep out of jail, and in his late teens and early twenties he set up his own car mechanic workshop. Over a few decades I did maintenance on my various cars at his workshops, as well as many of the modifications on my Volvo (to get it ready for extreme outback driving). He taught me a lot.

Unfortunately, his sister and brother couldn't break away from their tough start to life. Many years later, my wife,

Mary, and I were reading the newspaper in bed. The front-page story told of a young man with a female companion who had stolen a car and, while being chased by the police, crashed through a security fence onto the runway at Sydney airport after it had closed for the night. They raced down the runway, with a bunch of police cars in pursuit, and came to a screaming halt when the runway finished at the waters of Botany Bay. They were arrested, and it turned out it was my mechanic mate's brother and sister. I was simultaneously sad for them, but happy that it wasn't my mate.

Glebe today is a pretty upmarket suburb, very different from its tough working-class origins. I was fortunate that my mechanic mate took me under his wing when I moved into my first squat. He may have been younger than me, but he'd lived in Glebe his whole life and had street smarts beyond his years. Which was lucky, because I basically had none.

1975–83

Glebe squats

I moved through lots of different share houses as a hippie. I started off in inner Sydney, then out to bushland on the edges of town, then gradually back into the inner city. Sometimes households dissolved as couples formed (or broke up), but I always knew somebody in the house I moved into. Houses were more communal back then, and people in share houses spent time together at home.

Very few of my housemates had conventional day jobs. They made specialist electronic gear, super-large inflatable sculptures, movies, regular sculptures, music, poetry, paintings – a real cross-section of the arts scene in Sydney. At the least salubrious place I stayed in, my friend, the leaseholder, did so little housework that I actually pitched a tent in the bedroom just to find a clean patch to sleep. I didn't stay long at that one!

I moved into a Glebe share house around 1975 that, to my surprise, turned out to be a 'squat'. A squat is an empty house, unoccupied or even discarded by the owner. It is better than low-rent housing in that it is no-rent housing! Obviously there were still costs, like repairs (you couldn't exactly ask the landlord) and utility bills (however, there are ways to get electricity for free!).

It's odd but I can't even remember exactly how I came by the Glebe squat. I was invited in by somebody, and ended up staying in that same squat for about eight years. And over time – and I really, really apologise for this – I kind of inadvertently drove everybody else out of that squat! Not paying rent and electricity made it possible for me to afford to morph from being a roadie–taxi driver into a paid gig as a hospital scientist, and then later a long-term penniless student – first in engineering and then medicine!

The vast majority of people who moved into squats wanted to live there as long as possible, and were rarely party people who wanted to wreck the houses. But the squatters were definitely a mixed group. There were lots of 'regular' lovely people. There were impoverished families. There were injecting drug users. There were bikies. There were couples with or without babies. There were friends and single parents. Some were poor students, and some were runaways or unemployed, while a few had mental-health issues or were vulnerable in other ways. Many were there because they had no other option and were simply too poor to pay rent. And somehow (and I'm still not entirely sure how) there was single me. I was there because I was in transition – with a casual income and an as-yet-unknown destination. Overwhelmingly, my time in the squats was a fantastic period of time for me.

Today Glebe is a much sought-after inner Sydney suburb, so it is kind of hard to imagine that it ever had houses just sitting there unoccupied. It was always a highly populated working-class area, though, so how do you get squats in a densely populated suburb?

Well, a poorly conceived Department of Main Roads 'plan' involved deliberately encouraging the Glebe squats to come into existence. (Other states might want to contest this, but I've heard on the grapevine that New South Wales is consistently the most incompetent and corrupt state in Australia.) The idea, apparently, was to build a freeway from the Sydney Harbour Bridge to the Blue Mountains. The blueprint was just a few totally unsympathetic straight lines blasting west. To do this, the DMR was buying up buildings on the path of their planned western freeway route, starting with Glebe. They had an 'interesting' approach.

The Glebe houses initially bought by the DMR were dotted over a strip about 50 metres wide and several hundred metres long that covered half a dozen parallel streets. The houses were bought at reasonable prices from the owners, but as soon as the DMR took possession, they immediately removed the doors, leaving the houses wide open. The word in my taxi-driver world was that the DMR wanted to encourage 'riffraff' to move in, hoping this would bring down the tone of the street so then the remaining homeowners would sell up and get out quickly. A quick sale often meant accepting a cheaper price, especially if the area was seen to be rapidly running down. So the word on the street was that the DMR essentially created the Glebe squats, as a way to keep its budgets down.

Sure, this is a totally unproven conspiracy theory. On the other hand, the approach sure looked like a sneaky workaround intended to avoid paying market price for the remaining properties they wanted to acquire.

One building the DMR originally planned to obliterate was a historic residence in Darghan Street called Lyndhurst, just across the road from my squat. It was built around 1833, forty-five years after the First Fleet landed in Sydney Harbour, originally on some 1500 square metres of land. Very uncommonly for Sydney at that time, it was a rare colonial example with retained native shrubs and trees. The grand free-standing two-storey residence with wide verandahs, stables and service wings was designed by John Verge, an English architect and early settler.

Perhaps the DMR had some awareness of how bad it would look to unashamedly demolish a unique and historic house. But if the whole area became a slum, then maybe its heritage value would be less effective as an argument against the proposed trail of destruction through the suburb.

Thanks largely to local action groups, Lyndhurst was saved from the wrecking ball, although it was in a pretty poor state of decay by the time anyone got around to fixing it back up, with sheets of iron missing from the roof. Squatters had moved into Lyndhurst (as well as the surrounding terraces) and were using the grand old house in lots of different unsympathetic ways. One person had set up a semi-professional audio recording studio on the ground floor, bricking in a bunch of new walls – along with a touch of light demolition! He got his power from a 30-metre cable, strung through the air across the road to my place.

A year or so after I had moved into my squat, a wild windstorm hit Sydney, and several more sheets of iron tore off the roof of Lyndhurst and flew around Darghan Street. Afterwards, a neighbour and I wrote a succinct letter to the *Sydney Morning Herald*, which was published on the same day as photos of the storm damage around Sydney. (Maybe *this* was the start of my media career!) The letter was highly emotive and went something like: 'We the residents of Darghan Street, in Glebe, had to shelter in terror inside our houses as loose sheets of roofing iron were torn off historic Lyndhurst. One of these sheets could easily have decapitated any of the small children who were outside looking at the storm. Could the new state government please fix this important building before innocent children are killed?'

In May of 1976, the Labor Party had won the New South Wales state election with a paper-thin majority. At that time, the DMR plan was still for Lyndhurst and all the squats – including mine across the road – to be demolished. With such a narrow margin, the government was pretty sensitive to the need to raise its popularity, and pushing ahead with demolition of a heritage building against the wishes of locals would be a very bad look.

Whatever the prompt, within two days of our letter being published, a veritable army of workers descended on Lyndhurst and proceeded to spend a reputed $100,000 repairing it! That was a lot of money back then for any project, but this was a government that needed to buy some popularity in a hurry. Saving Lyndhurst from demolition and restoring it seemed to put paid to the rest of the freeway project. So I guess I played a small role, along with all the continued community actions, in preventing the destruction of historic Glebe.

My squat was in a row of identical small two-storey terraces and, like Lyndhurst, my terrace was going downhill pretty quickly when I moved in. It looked in a pretty squalid state even for a squat! I was genuinely worried that my housemates were going to run the place further down, or even deliberately vandalise it to the point of being no longer habitable.

I personally thought letting the house go to ruin was a bad thing, so I came up with a rather laborious plan to renovate it by removing all the wall plaster (also called 'scutching').

In Sydney back then, taking the plaster off old bricks was really in. It meant that people could see original brickwork that might be a century or more old – theoretically, a nice touch. So I bought a scutch hammer and a bunch of scutch combs and started chipping the plaster off the inside walls. What I didn't realise at the time was that the old brickwork of the 1800s was sometimes of really low quality. Occasionally there'd be a stretch where the bricks were the same size, evenly laid out and overlapping each other by half with the same thickness of mortar – no worries, nice and professional. But surprisingly, often there'd be entire bricks missing and replaced by half bricks, quarter bricks or even just crumbly mortar. Yep, it was actually quite ugly. But once I'd started, I felt I had to keep going – otherwise the place would end up looking even worse than before!

A problem that I uncovered during the scutching was that the walls separating the terraces from each other were single brick, not double, so the plaster had both an insulating and structural function. There was a little part of me getting nervous that I might in fact bring the whole house down with my amateur renovating.

The long-term plan was to eventually clean everything up and make the house cool and groovy. The newly naked bricks and mortar were continually shedding dust, so later I would seal them with transparent varnish. But in the short term, all the extra dust made living there very uncomfortable. I was creating infinite mess from when I awoke to when I went to bed. This became my full-time occupation.

The days turned into weeks, and in my single-mindedness, I continued chipping away the plaster from every bloody wall in every room, even if people were in them. I did not clean up the mess at the end of each day, and no one else lifted a finger to do so either. So in a surprisingly short time, in preference to being covered in plaster dust day and night, everyone else left. Nobody said anything to me – they just moved out. Finally, one day there was just me in the house, all by myself! It took a few days more to clean the mess.

My place (mine, all mine, and free of rent!) was a two-bedder on Darghan Street, Glebe, with a nice wrought-iron balcony upstairs. I was keen to fix it. The timbers on the verandah had been ripped out for firewood, so I replaced them.

I had a basic knowledge of electricity, so I was able to (but definitely not allowed to) put on some thick rubber gloves, bypass the meter and run 'free' electricity straight into my house. (Before I did this, there was electricity, but I never saw a bill.) A few months later, while I was reading on the balcony, I saw a pair of meter readers coming down the street.

I held my breath as they looked at my neat but totally illegal wiring, shook their heads in unison, looked up at me, waved and kept going.

There were three small rooms on each level and a tiny bathroom tacked on the back. To get to it you headed out the back door, down a narrow path between the terraces and under the open sky and possible rain — but that didn't seem too inconvenient back then. The bathroom had cold running water, a bath, a toilet and a light bulb. This was luxury compared to most other squats that had a toilet all the way back next to the rear lane. I bought a water heater and installed it along with two separate shower heads, so two people could shower at the same time (just in case a special friend were to pop by for the night). I thought this was living the dream. Other squatters would use my shower and toilet, but I was cool with that. I didn't want to put a lock on the bathroom, because the squats had a strong communal ethos.

I even had a small vegie patch and compost pile in the backyard. One day, when the circus came to Wentworth Park just at the bottom of the hill, I went down with my wheelbarrow and came back with a few loads of elephant manure — so light and airy. My vegies flourished for the next few months.

The simple kitchen had a fridge and a gas camp stove. I ate a basic diet, but it was pretty nutritious. For protein, I had soya beans, which I bought cheaply in 20-kilo bags from Chinatown. A pressure cooker reduced four hours of cooking to forty minutes and made my little gas bottle last much longer.

I had a thrifty trick for my fruit and vegies, too. Occasionally greengrocers would hail my cab to drive to Paddy's Markets where, around 1 am every Monday, big trucks loaded with fruit and vegies from all around Australia would turn up to sell directly to shop owners from all over Sydney. The greengrocers would load my taxi up with their purchases and off we'd go. (As an aside, the markets were home to rats as big as cats. Everybody gave them a wide berth, and pretended they were invisible!)

Based on this taxi knowledge, each Monday morning around 5 am, I would roll my wheelbarrow to the local greengrocer while they were changing over the old fruit and veg for the new batch. They happily gave me the boxes of old produce that they would otherwise have had to bin. About 90 per cent of my wheelbarrow load ended up in my compost, but I lived very well off the remaining 10 per cent – it might have been a bit long in the tooth, but it was perfectly edible. I steamed the vegies and added tahini for flavour. I still love soya beans and steamed vegies to this day, the prince of foods in my mind!

To round out my diet, I bought skim milk powder and combined it with water and a healthy load of Milo.

I was happy to have other people stay with me from time to time, but I preferred living there on my own. But even when I had the place to myself, things weren't always peaceful!

Bikie showdown

None of the people my mess drove out of the squat directly complained to me. But I came to find out that at least one of them was not happy with what I had done. Obviously, I had ruffled his feathers!

I was vaguely aware that, at the time, one of the ex-housemates was a very junior bikie (also known as a 'nom', short for nominee). Apparently, after he left, he had risen through the ranks!

A few years later, I was in regular employment at a hospital (but still squatting). One morning, I got up early for my usual run around Wentworth Park, something I loved. It turned out to be extremely lucky timing, because if I had woken a single minute later, it might have been too late to hear and see a gang of bikies rumbling around the corner.

Darghan Street wasn't somewhere bikies just turned up for fun, and they'd certainly never appeared en masse like this before, to my knowledge. It was a dead-end street, and it wasn't like they were collecting door-to-door for the Red Cross. As I cautiously stuck just the merest part of my head out the front door, I saw a long column of maybe twenty or thirty motorbikes still coming around the corner, straight for me! It was early, only about a quarter to six, and I had a fleeting thought that I had to be in scrubs by 7.20 am. But right now, that was the least of my problems!

Reflex panic kicked in. My mind went into overdrive. As soon as I saw those bikies, I knew they were coming for me, and why. I instantly figured that the former nom had by now gained some seniority in the gang and was coming back for revenge – along with plenty of his bikie mates. And I strongly suspected that apologies and the offer of a nice cup of tea and a bikkie wasn't going to calm them down!

I shut the front door and ran as quickly as I could out the back, down the lane and towards the corner. I ran until I was behind the bikies and out of their sight, but I could still see them. I heard the loud shout of, 'Hey, Karl, we're coming for you. We're going to kill you!' Some of them had already kicked the flimsy front door open and were fanning through my little house.

By good fortune, there was a public payphone outside the shop on the corner. From my hiding place, there I could still see and hear the bikies shouting their true feelings about me to the entire street. With my heart pounding, I called triple zero and asked for the police, saying, 'Come quick – there are bikies killing this guy at Darghan Street in the squats.' I hung up. It was close enough to true – they definitely *wanted* to kill somebody, and that somebody was me!

With zero delay, I called triple zero again and asked for the ambulance. When I got through, I screamed, 'Come quick! There's somebody being attacked by bikies and he's got broken legs and everything. Send an ambulance straightaway – the cops are already here.' And then I hung up.

Next up was the fire brigade, slightly changing the backstory to encourage a prompt arrival: 'There's a fire, and a dog is trapped, and there's all these injured people out the front. And the cops and ambos are here!'

Glebe is a very small suburb, and my place was about half a kilometre from the police and the fire stations. By sheer good fortune, the closest ambulance station was only a kilometre away. As I finished my last call, I heard distant sirens getting louder, and apparently from every direction. There were three different sirens, because, in those days, all the emergency services had their own wail.

From my position hiding behind the corner of a building, I could just see the bikies on the street looking at each other nervously. Luckily, the first vehicle to swing around the corner was a police car. As soon as it appeared, the bikies took off with almost military precision. The ambulance and fire brigade were hot on the heels of the police, and soon the street was packed with flashing vehicles, people in different uniforms, and my neighbours checking it all out. At that point it seemed like a bit of discretion and distance might be the best thing, so I hit the road, sticking with my original (but unavoidably delayed) plan to run around the park. In fact, I went for a really long run. I couldn't go straight to work because my car was parked out the front of my house, and I couldn't go back to the house until the coast was clear.

When I finally returned an hour or so later, the street was empty again. My front door had indeed been kicked down, and inside was turned upside down. If those were the worst of my woes, I could handle them!

My next-door neighbours passed on a message from the police, who said they'd like to talk to me. Yes, I imagined they might after all that.

I rang the hospital and told them I was running late because a neighbour had had an emergency that was now resolved, so I would come straight in and shower and change into my scrubs at work. Then I just drove to work like it was another ordinary day.

Of course, I didn't breathe a word about any of the morning's events to anyone at work. And I didn't ever call the cops back, and they never followed up – and luckily, the bikies never came back for me again. All that remained of the whole situation was the splintered wooden door, and that was an easy fix.

My house on fire

I settled back into happily living by myself and turning some of my more eccentric ideas into reality. I have a love of surrealism. One day, it came to me like a dream that the middle room downstairs would be perfect to affix a replica of the furniture that stood on the floor to the ceiling! So I mounted a standing light on the ceiling as well as a complete lounge, two sitting chairs and also (what the heck) a coffee table. Hoorah for long wood screws and a power drill! It was close to an exact mirror of the furniture on the floor. Even though the room was small, the ceilings were relatively high, so it totally worked for me, and other people said they loved it too.

After a few years, I got really sick of the naked scutched wall with misaligned and misshapen bricks staring at me. Besides looking bad, the multiple tiny holes in the mortar meant a lot of sound came through from the next-door neighbour's house. I could even see spots of light coming through from the other side of the wall, like a mini galaxy. Whenever they lit their fire, not only would a small amount of smoke come through from their side, but I could also glimpse a flickering yellow of the flames inside their fireplace.

I really hadn't thought it through at the time I was scutching. You see, when they made these terraces, they cut a few corners.

First, as I found out when I was doing the scutching, there was just a single layer of brick between one house and the next, not two layers.

And second, instead of using proper refractory fire bricks in the fireplaces that could withstand naked flame indefinitely, the builders used regular low-quality house bricks. And, like everywhere else in the house, not even whole bricks but sometimes half and quarter bricks, held together with lots of crappy mortar. And guess what? After a century of burning fires in the fireplace, the house bricks and mortar began to break down, which was why I could glimpse naked flames through the bigger holes.

My new plan was to fix the flames and the 'ugly' in one go by covering up the scutched wall with thin sheets of wood veneer. Amazingly, it worked – there was less sound from next door and no smell of smoke when they ran the fireplace. But one day, on their side, a great lump of mortar fell out of the fireplace wall. It made no real difference to them. But as you can imagine, the naked flames leapt through from my neighbour's fireplace onto my new wood veneer covering, and set the entire wall on fire.

Luckily, the neighbour was home and saw the smoke billowing from underneath my front door. They rang the fire brigade immediately, who of course smashed through my flimsy front door (again) and put out the fire. My neighbour phoned afterwards to inform me that my house was on fire, but nobody got hurt and the fire brigade put it out. What the heck?!

I realised pretty much straightaway that using flammable wall coverings anywhere near naked flames was bound to end in tears. The whole place nearly burned down, and it was totally my fault! I took down all the burned remnants of timber veneer and organised a council pickup. Luckily, the fire hadn't had time to spread to the floorboards or anything structural.

It was a few days' work to make it liveable again, this time with more forethought about noise, smoke and flames. (I have had a sudden realisation that these stories show me to have more than a touch of Homer Simpson's boundless and enthusiastic ineptitude!)

In general, I had good relationships with all of the people in the squats, especially the couples, families and single parents. A lot of us were poor, but here we could live with dignity because we didn't pay any rent. For about five of the years I spent there I had no income, first as an engineering student and then as a medical student.

For a little while, the terrace right next to me was a bit of a dosshouse, with an accompanying scene of drugs being sold and fights breaking out. I was robbed once, and even though I had no proof, I was pretty convinced it was their doing.

But to be fair, we were all just trying to survive.

Martial arts save the day

The tipping point came when one day I arrived home to find my lovely toilet seat torn off its hinges and used syringes scattered over the bathroom floor. The toilet was outside the main house, in a shabbily constructed lean-to. So anybody in my block of terraces could come out of their back door and use it. There were no fences separating the backyards – they had long ago been burnt for firewood.

I definitely didn't want a needle-stick injury in the dark, in my own bathroom.

My instinct was to replace the long-gone fence to protect my bathroom – but it wasn't quite in the spirit of the squats to wall off a common access area like the yard. So I had to go slowly. I didn't want to put people offside by behaving like a greedy developer.

I started off by planting more vegetables in a rectangular patch that extended from the back bathroom wall most of the way to the rear lane. And naturally, I planted my favourite 'Purple King' climbing beans. Later, when the lovely little beans began to climb, I naturally needed to add a wire mesh for them to hang on to. Which naturally meant I had to put in some vertical posts with wire mesh stapled on. And what do you know, the whole setup 'accidentally' acted as a complete wire enclosure around my now private backyard.

But I didn't take the garden all the way to the back lane. I deliberately left enough room for people to walk through from one backyard to another, behind my garden. One day I came home from hospital to find that all my hard work had been destroyed. The entire garden fence was torn down, my beloved climbing beans had been trampled into the ground, and the unflushed toilet was filled with faeces. It was pretty miserable.

I knew then that it was time to disorient the opposition and completely change tack.

Luckily, I had read *The Godfather*. It was my source of moral and spiritual guidance – much more useful than the Bible. To me, it was a guidebook on how to survive and prosper in a hostile environment.

One of the messages that really stuck with me from *The Godfather* was to never make a threat unless you're prepared to carry it out immediately. So when one of my difficult neighbours yelled, 'I'm gonna kill you,' I no longer kept quiet but just replied, 'Fine, here I am. Do it right now.' He looked startled and left, muttering, so I knew then that he was just exercising his throat muscles and was no real physical threat.

After the latest bathroom incident, I decided it was time to make a show and stand my ground. I changed my afternoon exercise habits so whenever I could come home early, I took to ostentatiously pumping iron in the backyard where everyone could see. I even bought some weights from one of the people in the house next door, the ones whom my bravado was aimed at, which just goes to show how weird personal relationships can be. I also practised martial arts in the backyard Bruce Lee style, Jeet Kune Do. I was training in martial arts classes at the University of New South Wales, so I knew some routines. I 'borrowed' some hospital theatre scrubs and dyed them black. I painted a broom handle black, then carved some random kanji characters into one end, going over them in gold Texta – my very own fighting stick! And there I was, pumping iron and doing my martial arts routines. I was definitely very amateurish, but at the same time, I was far more expert than anybody else in the squats.

I spent an hour each day going through my routines, making a point of inventing new routines when the antagonistic neighbours walked past. They mocked me at first, but after a while they just watched. Two months later I rebuilt the fence for the garden. It never got pulled down again.

Cops come calling

There was another occasion when I left my squat at exactly the right time, though this one was a little less dramatic than when the bikies dropped over uninvited. This time, I was just out the door and heading down the street for a run when I saw the cops driving along my road in my direction.

I had a bit of a chequered driving history from my taxi days, with a whole lot of unpaid fines floating around with my name on them. Every now and then, the thought crossed my mind that some chickens might one day come home to roost. My Spidey sense was tingling! Yep – this was almost certainly that day. Thanks to my previous practice run-through with the bikies, I knew exactly how to handle this situation. So I kept on running down the street, nice and calmly, and stayed out for a very long time.

The cops left a card on my front door saying there were a bunch of tickets in my name for unpaid speeding fines – no surprises there. Could I please come to the Glebe police station to chat about it?

By an amazing coincidence, I was heading on a road trip to Tasmania with my parents that very day. So after work, I drove south to Wollongong, stayed the night with my parents, then spent a few days driving down the lovely coastline to Melbourne. Then we jumped onto a ship, crossed Bass Strait and drove around Tasmania for a few weeks.

Finally back on the mainland and heading home, I asked my father what I should do about the police who wanted to 'talk' to me. My father, with his legal mind and training, knew just what to do.

'Throw yourself on the mercy of the court and take the chequebook with you,' he said. 'Just make sure you've got enough money to cover all the fines. And make sure you're well-dressed, and carry some kind of official badge related to the hospital.' I was cool on the money front, because I had a hospital job at this stage and a bit of savings.

When I finally arrived home on the Sunday night after the holiday, there were an awful lot of calling cards from the Glebe cops! I went back to work at the hospital on Monday, and then straight after work, I fronted up to the cop shop.

Just like my father suggested, I was wearing my best suit (I only had one, a safari suit, the pinnacle of fashion) and had my hospital ID prominently pinned to the front of my coat. Dad had also told me to put my pager in obvious view. (In those days before cellular phones, I always had to carry a pager. Whenever I heard it beep, I would immediately find the nearest public phone and ring the hospital for further instructions.)

I walked up to the sergeant at the front desk. 'G'day. Look, sorry, but I've just been in Tasmania for a few weeks and found all this stuff in my letterbox, so I came straight up.' When I handed across all the police calling cards, he asked how long I'd been away. Luckily I had my boat tickets for the ferry across Bass Strait in both directions that showed I had landed back just one day before. (Thank you for your detailed advice, Father!)

The police officer glanced at my ID and the beeper bulging out of my safari coat pocket and said, 'What's that?'

I told him I was a scientific officer in the cardiac catheter laboratory at the Prince Henry Hospital (hoping that made me sound like a Very Important Person) and that's my pager, because I'm on call all the time for medical emergencies. Then I apologised again for missing their visits and asked what it was all about.

'Looks like you've got some unpaid speeding fines,' he said.

'Well, look,' I said, 'I used to be a taxi driver, and you know what it's like being on the road. If I did some bad things, then I've got to pay the fines. Luckily, I've got my chequebook with me, so would you like me to pay right now?'

The cop was so happy that I just wanted to pay the fines right there on the spot. Sure, there was some paperwork for him, but it was the bare minimum. And definitely no going to court for either of us.

To top it all off and show what a lucky bloke I am, they seemed to have lost not just some, but most of, the fines. I only had to pay about a fifth of what I expected to pay, a few hundred dollars, and once I wrote the cheque I was totally off the hook, with an unblemished future to look forward to. And best of all, no more unexpected early morning wake-up calls for me!

It really could have been so very much worse. If I had not luckily transformed myself into a Pillar of Society, with a proper job at a hospital and a beeper and a chequebook, I could have easily ended up sharing a cell with a large and hairy man who had LOVE and HATE tattooed on his knuckles.

Getting straight

Becoming a hospital scientific officer meant I permanently left taxi driving behind. I really hadn't driven much at all after moving into the squats, because I didn't need much cash.

In the spirit of going straight, I decided to try to formalise my housing situation with the DMR. I wrote a letter saying that I had been living at the house on Darghan Street, Glebe, for some time, had improved it and kept it in good condition, and I would now like to set up some kind of formal agreement about residing there. What would they suggest?

A response promptly arrived on official DMR letterhead saying that they had received my letter and were considering it. It went on to say that in the fullness of time, and after due process, they would let me know their decision. The very next day I received a second letter thanking me for my first letter and further acknowledging only one thing – that they had sent me a previous letter! There was no more correspondence for several years. I idly wondered what would have happened if I'd replied to their reply. (I still admire the DMR for its perfect *Yes Minister*-style bureaucracy.)

The good thing was that now I had a document on government letterhead, to me, at my actual address. It was my first 'official' recognition as a resident of Darghan Street. Now I could apply to get a landline phone in the house, because I was somewhat sanctioned to live in the Glebe squat! My first phone was a lovely red Ericofon, which I still have working today. To me it's an item of classic design! To keep the Ericofon running I had to convert it from analog to digital, which meant a whole special converter box. I admire it every day I walk past it. Nobody rings me on the landline anymore, but it's fun to show my little nieces how we used to dial by putting our fingers in the holes matching the telephone number.

That lovely little squat, with its free rent and free electricity, was my beloved home for eight years. After my father died a few years later, my mother sold up in Wollongong and bought a place in nearby Lilyfield that had two houses squashed onto the same small block. I didn't want her to be alone, so I moved in with her and passed 'tenancy' of my squat on to a mate. After a few years, the state government decided to take the squats back and made him an offer he couldn't refuse: modern subsidised government housing nearby.

Later the government sold the house and it became a private residence again. Farewell, my lovely, we had good times ...

1976–77

Hooroo, hippie! Hello, hospital!

In my late twenties, after several years as a hippie, my lifestyle changed dramatically. It wasn't quite overnight, but landing a job as a scientific officer in a cardiac catheter laboratory led to a total U-turn in regards to my personal priorities. I was working at the now-closed Prince Henry Hospital at Little Bay, in Sydney's east. Suddenly I was fronting up to hospital five days a week and doing on-call shifts as well. I spent two years in this public hospital role, thinking of myself as an agent of the government, because I was paid from the public purse, while also trying to be an agent for change and enlightenment.

As it turned out, if I hadn't been a hippie who knew how to build a film-processing laboratory, I wouldn't have been able to become a scientific officer! And although I didn't know it at the time, my years as a scientific officer would lay the groundwork for my future careers as a medical doctor and Champion of Science. Each 'career' was an accidental stepping stone for the next one.

Serendipitous lucky breaks have been woven throughout my life, always completely unplanned and unexpected. I guess the message is that learning is never a waste of time, because you just don't know what could be useful for you down the track!

Hippie to hospital?

One big motivation to move into a professional role was getting paid a measly forty dollars in my hot little hand, after I put so much time, money and creativity into making the music video a few years earlier. Let me emphasise that it wasn't just the pitiful sum of money – it was the complete lack of recognition of all I had done, and the lack of respect for my craft and my time. It still rankled with me that I got so little, for so much effort!

I was really disillusioned in terms of the old filmmaking dreams, and I didn't have the resources or the foresight to stick around until music videos became huge, which was actually not that much further down the track. As a hippie, I had tried to forge my own destiny. But after the five years covering 1971 to 1975, I still didn't have a reliable income and was living in a squat! Yes, my life was a little messy when you looked at it from certain points of view (such as my parents'!). I personally loved living in the squats, but I can see how some folk would see it as a low point in one's life trajectory.

My parents were rightly worried about where I would end up, if I didn't smarten up. They wanted me back on the straight and narrow. One weekend at their place in Wollongong there was a discussion about a job at a hospital in Sydney, and with a strange combination of reluctance and enthusiasm, I applied.

On the one hand, 'working for The Man' felt like a betrayal of my hippie dreams, plus I really did like being able to sleep in anytime I wanted. But on the other hand, there was my father, who was ready to take control (in retrospect, thank you sooooo much, Dad).

My father helped me 'massage' the last five years of my life into a successful job application. The next thing he did in making me employable again was to take me directly to the dentist! (Dental hygiene was not one of my priorities as a hippie, but oh boy, how things have changed!)

The dentist worked carefully through a long list of necessary fillings and cleaned my crooked teeth. He also taught me the correct way to use a toothbrush. Most excitingly, he showed me how to use something I had never heard of previously, but now love to pieces – dental floss. (Years later, as a junior hospital doctor, I was admitting a spritely ninety-year-old woman. As part of the physical examination, I saw that her teeth were all there and in perfect condition. This was really unusual for a sixty-year-old, much less a ninety-year-old. She told me that for as long as she could remember, she had used a silk thread to clean between her teeth.)

They say the suit maketh the man, and my dad agreed. Back in the late 1970s, the height of sartorial elegance was the light brown safari suit. For those who have never had the pleasure of wearing one themselves, a safari suit is a combination of a lightweight short-sleeved jacket – with its own belt, epaulets (!) and four (or more!!) expandable (!!!) pockets – and pants. So my father took me to downtown Wollongong and got me fitted up. Dad chose the jacket and trousers combo (rather than the shorts) because he thought it was more 'formal'.

Dad finished my makeover by getting me to his barber for a neat short haircut and a clean shave. My freak flag was gone. I didn't weep openly, but inside, tears were flooding silently from my heart. However, I succumbed willingly to my dad's advice, because I knew he was right. Then he happily sent me off to the interview, having achieved his simple goal of making sure I no longer looked like a drug-crazed hippie!

The interview was interesting.

Instead of chatting about my hippie years, I concentrated on explaining how my degree in physics was actually a degree in problem-solving, which meant that once I had picked up a bit of specific knowledge, I could do any job. (If it worked at the steelworks, why not use the same line again?) It must have been convincing, because they didn't seem to mind that I had a complete lack of knowledge of any health science, and zero hospital experience! I also redefined my taxi-driving and roadie phases (and totally left out the free-love hippie vibes!) as reflections of my independent and creative spirit.

Then came the filmmaking. Everything suddenly changed.

Weirdly, the interview panel seemed completely focused on my having built a complete 16-millimetre colour film-processing laboratory. But that very soon made sense, when they told me their cardiac catheter lab shot and processed several reels of 16-millimetre black-and-white film every day. (Back then, domestic video tape simply didn't have the definition and contrast that film had. And colour film gave them no advantage.) I slowly realised as they described the job to me that the single goal of the entire catheter team seemed to be to capture some kind of essential diagnostic information on movie film. That information would determine the next step in the patient's care.

The bottom line seemed to be that my film 'career' combined with physics was exactly what they wanted.

When I relayed to my father the interview discussion, he listened very carefully and asked pointed questions. At the end, he confidently told me, 'You've got the job' – and he was right!

Hippie vs hospital

Within a few weeks of my interview, I was a real hospital scientific officer, complete with colour photo ID badge. And despite my trepidations, I was loving having a regular wage, with the added bonuses of holiday pay and sick leave!

In my years as a taxi driver, I had been beaten unconscious once, scared for my life more than once and mourned two of my driver mates who had been killed while working. I was pretty sure that the hospital would be a much safer workplace.

My first week as a scientific officer (SO) was a shock to the system, but also lovely. I got thrown into the deep end, and it turned out to be a great pool to swim in.

I was the most recent and inexperienced addition to the team, so I kept my big mouth shut and my big ears open.

I obviously had to be honest with the nurses and doctors about my total lack of knowledge in cardiovascular 'stuff', but they were very kind to me. In the quiet times between patients, they would teach me the anatomy, physiology, histology, electrodynamics and more that I needed to know about the heart. They also lent me relevant books to read. I was trying to get up to speed as quickly as possible so I could do my job independently without the other SO having to carry me.

As a taxi driver, I had been proud to provide a service to society, but working in a hospital took service to a whole new level. At the hospital, the overall daily goal of most staff was to help their patients live well. I felt so happy to be part of a team that was working to make society better. I really didn't anticipate the sheer pleasure I would receive from being able to help each patient, one at a time.

In the hour or so I spent with each patient, I tried my hardest to make sure they had the best possible experience in that time. Part of my role was to meet them on the trolley, explain to them what would happen in the operating theatre, and what would happen afterwards. During the catheterisation procedure, which was filmed, my role was to take and record accurate measurements related to their cardiovascular system (for the medical team). Afterwards, I would process the film.

```
            Heart 101: Pumping blood

Understanding how a heart should work (but
usually didn't in our patients) made me fall
in love with the human heart. So let me
```

share some of this awe and wonder with you.

Your heart is the size of your clenched fist and weighs about 200 grams. Regardless of its tiny size, over the two million seconds of your life it will pump around a million times its own weight. That's 200,000 tonnes of blood, roughly the weight of one of those ridiculously large cruise ships.

The heart is divided internally into four separate chambers (or little bags). Around the outside of the heart chambers are various muscles. In a carefully timed sequence, the muscles contract in a beautifully coordinated manner, increasing the pressure in each chamber to pump the blood inside to its next destination.

Deoxygenated blood flows at low pressure from a large vein called the vena cava into the right atrium, then into the right ventricle, and is then squirted out into the thin walls of the lungs' blood capillaries. These capillaries are cleverly folded to have a total surface area of about 70 square metres – a bit less than half a tennis singles court (thank you, Evolution!).

In the lungs, carbon dioxide leaves the blood while oxygen flows in. The now-oxygenated blood moves from the lungs into the left atrium, then into the left ventricle, and finally, out through the aorta. The aorta is huge – about 2 to 3 centimetres in internal diameter, or roughly the thickness of your thumb.

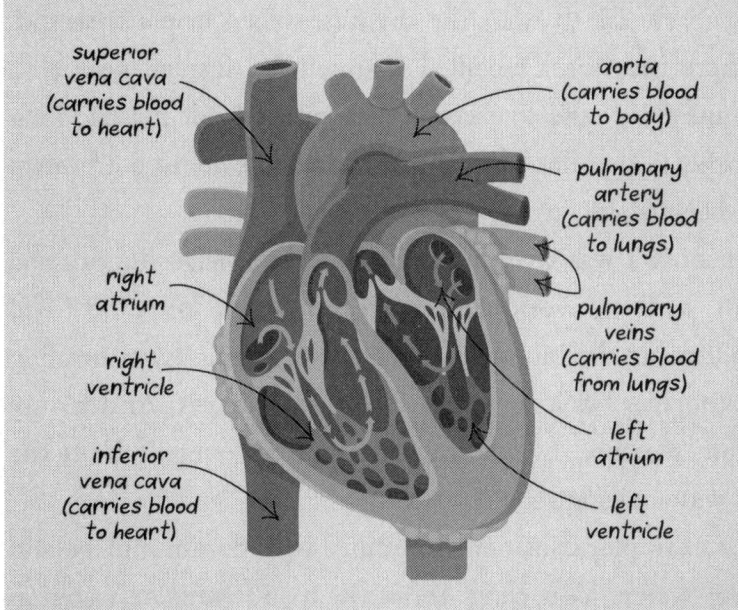

superior vena cava (carries blood to heart)

aorta (carries blood to body)

pulmonary artery (carries blood to lungs)

right atrium

pulmonary veins (carries blood from lungs)

right ventricle

left atrium

inferior vena cava (carries blood to heart)

left ventricle

(Just as an aside, the air entering your lungs has around 21 per cent oxygen, but when you exhale it's only about 17 per cent oxygen and 4 per cent carbon dioxide. The rest is practically all nitrogen.)

The end result is that every second or so, about 80 millilitres of high-pressure oxygenated blood is pushed into your general circulation through the aorta, adding up to about five litres of blood each minute.

The biggest intellectual surprise was opening my eyes to the astonishing universe inside our bodies.

As a physicist, I honestly had no real knowledge about the internals of the body that I walked around in. I had a rough idea that my body was filled with blood and a few lumpy bits called 'organs'. I could name the heart, lungs and liver, but I had no clue as to what the thymus, spleen or pancreas did, or where they sat. Overall, I assumed my insides were akin to chunky red salsa that could leak out if I cut myself. That was close to the sum total of my knowledge about the human body back then.

But now I was working in a scientific role in the hospital system, with my work centred on the human heart. So I had to understand the biology of what I was doing. I was heading from the Hard Sciences into the Wet Sciences. And to my absolute pleasure, every single thing I discovered about the body was amazing!

For example, consider something very commonplace and boring: water. You carry about 40 to 50 litres of water in your body, but each day a total of about 40,000 litres of water crosses the many membranes inside your body. That's because each molecule of water is constantly shifting, across membranes and back again, about a thousand times each day. That's roughly the volume of a backyard swimming pool! (And it weighs about 40 tonnes). The outside of your body has a surface layer of stable skin, but inside you are a boiling maelstrom of watery commotion! And that's just boring old water, not even one of the many magnificent organs, such as the heart.

Heart 102: Absolutely essential arteries

The muscles around the chambers of the heart (the atria and ventricles) start pumping around the third week after conception. After that, they keep on pumping until you die. They never stop.

These muscles need a constant supply of fresh oxygenated blood to keep working. They get this from two arteries (the right and left coronary arteries) that branch out from the base of your aorta. An artery is a pipe that carries high-pressure oxygenated blood to the various systems in the body. (A vein is a similar pipe, but it carries low-pressure deoxygenated blood back to the heart.) These coronary arteries are the very first arteries to get access to the high-pressure oxygenated blood coming out of the left ventricle.

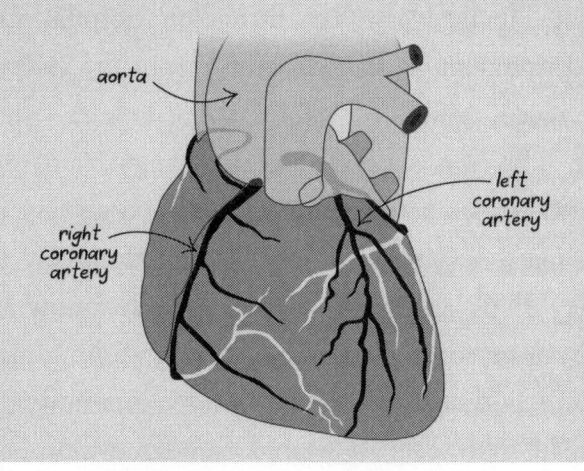

> These arteries (and especially the left coronary artery) are some of the most essential arteries in the entire body. If the artery to an arm or leg gets blocked, that arm or leg will die – but you can still live. But if the left coronary artery gets blocked, the muscles on the left side of the heart can't contract properly and you might have a heart attack and die.
>
> Our daily bread-and-butter in the cardiac catheter lab was imaging these coronary arteries, and then measuring blockages within them.

Fringe health benefits

Totally unexpectedly, my diet shifted dramatically when I began work at the hospital and my nutrition became much more diverse.

I bought lunch every day from the hospital cafeteria (which had uninterrupted ocean views from the café tables). Fish, meat, salad, grains, vegetables, soup, dessert – they had everything. This was a very pleasant change from my soya beans, brown rice and vegies diet of the previous few years.

As a hippie, I had become more distant from my parents because they didn't approve of my lifestyle. But now I began to see them more regularly. I didn't have a washing machine, which hadn't bothered me much before, but now I was a working man who needed freshly laundered clothes every day. My visits home to see my parents in Wollongong were fortuitously timed to be around when I ran out of clean clothes, at which point my mum swung into full mama-bear mode.

Not only did she very kindly do my washing, but she also started organising my breakfast by modifying a traditional Eastern European dish called kutia and making large bowls of it for me to take back. Kutia (or kutya) is a sweet grain-based dish that includes wheat, poppy seeds, various nuts, raisins and honey, and is granular and nicely chewy (in my mother's version). It's popular in Poland, Ukraine, Russia and Belarus, but usually eaten only during the Christmas period. But if you were creative, it also made a tasty, nutritious and filling breakfast!

Dinner back at the Glebe squat was still good old soya beans and vegies with tahini, but now there was fruit too! At the age of twenty-eight I discovered a mango in the fruit shop on the way home from work. For me, it was 'Heaven on a stick'! I introduced mangoes to my parents, and when they were in season, I'd bring them a case on the weekends. Mangoes need to be eaten with your whole face, so I loved to eat them in the bath!

Physically, I felt great. I had started training for and running in the annual 14-kilometre City to Surf race (my best time was about 60 minutes). I would sometimes run along Anzac Parade for exercise if the operating theatre was quiet.

Finally, also at the age of twenty-eight, I learned to swim properly. It was life-changing. Somehow I had grown up and gone to school near the beach in Wollongong and never learned to swim, and it was always something that held me back. I signed up for adult swimming classes and felt truly liberated and, finally, fully Australian!

My new job was located a few hundred metres from the beach at Little Bay, and swimming at lunchtime became an absolute pleasure for me. Sometimes I'd run down to Little Bay and swim the 100 metres across the bay and another 100 metres back again. Then I'd run back to hospital, shower, change into scrubs and be ready for the next patient.

One day, when I got out of the water, I was surprised to see people up on the hill frantically waving at me. I ran up to them and they told me in a panic that, as I started swimming across the bay, a huge shark (at least half as big again as me) came out of nowhere at high speed. They watched in dread as it slowed down and swam just a few metres off to one side, closely tracking me all the way across the bay. When I did a U-turn to swim back, it lost interest and headed out to sea again! I had no idea it had been there, thankfully. (I almost had a heart attack just hearing about it, but it would have been poor form to be the patient in my hospital, not the scientific officer!) And despite this close encounter, I kept on swimming whenever I had the chance.

Unfortunately, in 2022, a swimmer was attacked and killed by a great white shark in Little Bay, the first fatal Sydney shark attack in six decades. There were people fishing nearby, and it's thought the fish bait in the water attracted the shark. Overall, the general fear of sharks is totally unjustified given how few people actually die from shark bites. Averaged over the last five years, the number of unprovoked shark attacks across the whole world is around seventy, but only about five to ten deaths happen each year. (In 2020, there were at least a hundred deaths from ladders in the United States alone, so being strictly rational, we should be more scared of ladders than sharks! When will the Ladder Lifeguards be formed to protect Australians from the dark, lurking threat of death by ladder?)

As the old saying goes, the only two things you can't avoid are death and taxes. I survived the shark, but the taxes were coming to get me!

It was now 1976, but I'd lived in a strictly cash economy for the last five years, which made it easy to not pay taxes. But now the hospital pay office wanted my tax file number. What to do?

From Hollywood movies I had learned that, as a rule, you never mess with the tax department. I knew that Al 'Scarface' Capone was a big criminal in Chicago in the 1920s, but it wasn't his violent crimes that ended up putting him in jail. No, it was his tax evasion!

Once more, my dad stepped up as my 'fixer'. He drafted a letter on my behalf to the tax office saying that I had been a taxi driver and penniless filmmaker since 1971, but now that I had a 'proper job as a Hospital Scientific Officer, with the responsibility of being on call for Medical Emergencies' (exact quote), I wanted to formally re-register myself with the Glorious Australian Tax Office! As to being 'off the books' for the last half-decade, well, I was now simply 'throwing myself on the mercy of the court' (my father, the ex-lawyer, insisted on using that exact phrase) for not having done the paperwork and humbly 'awaited their decision'.

Shortly after, I got a reply reprimanding me for being so slack. They also warned me that if I ever earnt less than the taxable income limit in the future, that I had to (at the very least) fill out a non-lodgement advice. And as a bonus, there was no back tax owing.

Hooray! I was back in The System, *and* with a clean slate.

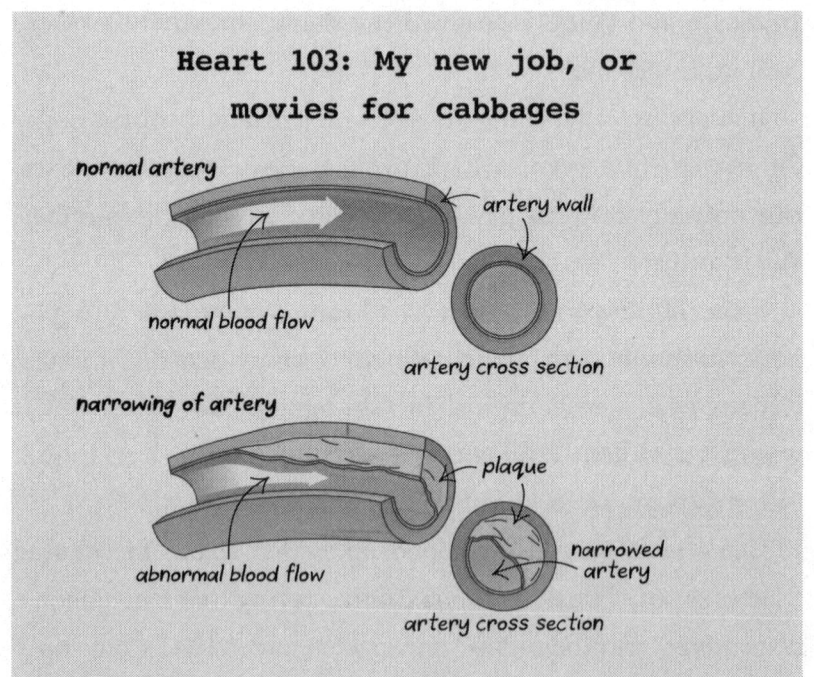

> ### Heart 103: My new job, or movies for cabbages
>
> *normal artery* — *artery wall* — *normal blood flow* — *artery cross section*
>
> *narrowing of artery* — *plaque* — *abnormal blood flow* — *narrowed artery* — *artery cross section*
>
> In the lab, our most common job was to measure blockages inside the left coronary artery. Such a blockage was called an 'atherosclerosis', or a fatty growth. These growths effectively reduced the diameter of the artery.
>
> The cold equations of the Hagen-Poiseuille Law of Hydrodynamics mean that the volume

of flow in a pipe (such as an artery) is proportional to the 'fourth power' of the diameter. This is bad news in terms of blood supply in a narrowed artery. Suppose that a blockage reduces the diameter of an artery by two. The flow does not just reduce by two times or down to a 50 per cent flow. No, the Hagen-Poiseuille law says it reduces by 2 x 2 x 2 x 2, by the 'fourth power', which is sixteen times. This means the flow plummets from 100 per cent to 6 per cent! This drastic drop in blood flow drives you into dangerous heart attack territory.

If we found a severe blockage and the patient was suitable for an operation, the doctor and patient would discuss the surgical options at a later follow-up appointment. If and when they went ahead, the heart surgeons would cut a vein from the leg, open up the chest and locate the blocked section of the coronary artery. They would then graft in the leg vein section to bypass that blocked coronary artery.

The procedure was known as a Coronary Artery Bypass Graft, or CABG, which, being Australian, we pronounced as cabbage.

My job was to make movies of blocked blood vessels so that the surgeons could do cabbages!

More changes in the background

I loved my Chevy and drove it to work every day. Most weekends, I would drive my beloved American Iron down to Wollongong to hang out with my parents. I felt closer to them, but we still disagreed about some things.

Usually, while I was visiting them I spent time working on my car (after all, it was practically vintage). My parents continually pointed out that this was a waste of my time, and I would counter this by telling them I loved doing car maintenance, and I loved the power of my Chevy.

One weekend, a friend of my parents turned up in her 1974 Mitsubishi Lancer. It was a rather small car, especially when compared to my giant Chevrolet. My parents told me that their friend was selling it because her husband had died, and that it had very few kilometres on it. Would I like to take it for a test drive?

I wanted to say no to the test drive, because I loved everything about my 1950s auto, but I agreed because, deep down, I knew that the Chevy was indeed taking up a lot of time in my life. So I gave the Lancer the standard Wollongong hot-rod 'grunt' test by checking out how fast it could go up Mount Ousley, one of the two ways to get to the tableland behind Wollongong. Part of me wanted to not like this car, so I got my parents and the seller to come with me to add extra weight and make it harder for the little engine. To my total surprise, we pulled 110 kilometres uphill on Ousley – it was about the same speed as my beloved Chevy! It also had better noise and vibration properties, and better and newer mechanical and electrical technology (disc brakes, anti-intrusion crash bars in each door, transistor radio, etc.).

When my parents asked what I thought, I simply replied, 'Well, it's actually a good little car with a bit of grunt.' That was the end of the conversation and no more was spoken of the Lancer.

When I turned up next weekend, the Lancer was sitting there in the driveway – my parents had bought it for me! They had never owned a car themselves in Australia and deliberately chose to travel overseas rather than spending money on one. My father could drive, but he never got an Australian driver's licence. They went everywhere on public transport. It was both a wise and generous decision on their part, and so I went from driving American Iron to Japanese Buzzbox, and it was great.

I gave my Chevrolet away to a friend. It had been good to me, but it was time to move on.

So now I had a reliable car that was small and pacy but still powerful enough to be fun. It was such a nicely balanced package of a car, and I later found out that it was successfully used as a rally car.

I ended up really loving the Lancer. An artist friend once drew a picture of it for me, complete with the hessian water bag that I used to hang off the front bumper bar on trips out of town so I had cold water anytime I stopped (the water slowly drips out and evaporates off the surface of the bag, keeping the water remaining inside the bag cold). I kept that Mitsubishi Lancer for about a quarter of a century and quarter of a million kilometres! That was a lot for a car back then – I put this down to my obsessive oil changes every 1000 kilometres. I didn't give up tinkering, though – I still did all the maintenance and repairs myself.

My hospital job in detail

Each day at the lab, we would diagnose some kind of heart disease in up to half a dozen patients using a cardiac catheter. They were mostly middle-aged and older, but there was the occasional younger person and sometimes even babies.

'Cardiac' is a fancy word meaning 'heart', while 'catheter' literally means a pipe. In medicine, catheters are used to get to different locations inside the body. Once there, you can either deliver some liquid, or take a sample of some liquid, or simply make observations, such as measuring pressure. Catheters are used for both diagnosis and treatment.

First cardiac catheter in humans

The very first cardiac catheter (guiding a pipe into a human heart) happened illegally in Germany in 1929, before there were any 'bothersome' routine ethics approvals.

A junior surgical resident named Werner Forssmann had the brilliant idea to thread a thin catheter through blood vessels and into the heart. He then planned to inject a radio-opaque dye into the catheter, so it would show up on an X-ray. If you took

X-rays at the same time as you injected the dye, then you could see where, and on what timescale, the blood flowed inside the heart.

Nobody had ever done this on a human before (because of possible serious side effects, like bleeding, stroke, heart attack and death!). But on the other hand, Forssmann thought that if it could be done without killing the patient, the test results would be very useful for both the patient's diagnosis, and to further scientific knowledge. He was even willing to perform the procedure on himself first.

In medicine, doctors have a long history of doing experiments on themselves, but in this case, Forssmann's boss was dead against his proposal. Time for Plan B – do it, but without his boss's knowledge.

His next move was to charmingly convince the operating room nurse, Gerda Ditzen, that his idea would be a terrific advance in patient care. (He neglected to tell her that his boss had absolutely vetoed his project.) Forssmann needed her onside because she was in charge of the medical instruments essential for the catheterisation. It took two solid weeks, but eventually she agreed to help him, though with the very specific condition that he trial this risky procedure on her and not himself. We don't know why she insisted on this condition. It seems so very selfless and brave.

Forssmann agreed – but he lied.

Together they laid out all the medical instruments, then Ditzen lay down on the operating theatre table. He strapped down her legs and arms, which was standard procedure for an operation in those days, then injected local anaesthetic into her left arm around the front of the elbow – but he also injected it into his own left elbow.

Ditzen couldn't see what he was doing as Forssmann proceeded to cut the flesh on the front of his left elbow with a scalpel and then push a urinary catheter (cardiac catheters had, of course, not yet been invented) into his own vein. Then he pushed it another 30 centimetres. Once he figured the tip was somewhere around his left shoulder, he covered the wound site in his elbow and turned his attention to Ditzen.

By this stage she understood he was experimenting on himself, and not her, as they had previously agreed. He managed to calm her down, unstrapped her from the table, and convinced her to chaperone him down a flight of stairs to the X-ray room. Together they switched on the X-ray fluoroscope screen – jackpot! They could see that the radio-opaque catheter was now in his shoulder region. As the X-rays continued to spray into his chest, he gradually pushed the catheter further in, about another 30 centimetres. He saw the tip of the catheter turn gently as it headed downwards towards his heart. He then pushed it down through the 7-centimetre-diameter superior vena cava, into the right atrium until it lodged inside the right ventricle of his heart.

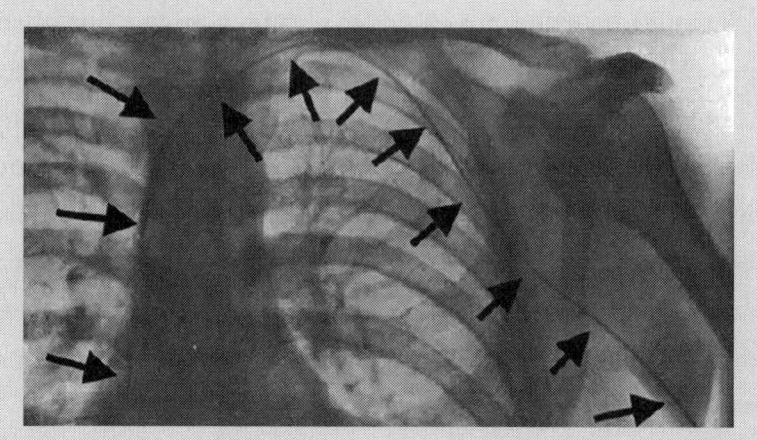

Original 1929 X-ray photo of the first cardiac catheterisation. (Arrows point to thin catheter coming from left elbow to right side of heart.)

(Source: Werner Forssmann, 'Die Sondierung des rechten Herzens' ['Probing of the right heart'], *Klinische Wochenschrift*, Vol. 8, No. 45, 1929)

> Success – a world first! Even better, he was still alive! The X-ray people took the very historic photo.
>
> His boss was remarkably unimpressed by his groundbreaking (and rule-breaking) research, and Forssmann subsequently had a very rocky career. It wasn't until 1956 that he was finally recognised and received the Nobel Prize for his pioneering work.

Unlike Werner Forssman, I kept my job strictly routine. I wasn't trying to rock the boat. Not just yet anyway.

I would help the nurses get the patient in and out of the operating theatre. There was a heated box with blankets inside – I just loved asking the patients if they wanted a warmed blanket! The patients were usually fully conscious throughout the procedure as it was done with a local anaesthetic. I also loved talking with the patients while they were waiting in the corridor, as the medical and nursing folk were preoccupied with scrubbing up, and I always tried to answer their questions and reassure them.

I learned all about the concept of a 'sterile field', which meant zero (or virtually zero) bacteria and other germs. The one or two medical doctors working on the patient would be scrubbed up in full sterile operating gear, along with the two or more theatre nurses. The scientific officer's lab came off the main operating theatre, so I had to walk through the operating theatre – but I usually wasn't scrubbed in with the surgical team. As a scientific officer, I wore non-sterile scrubs, booties, hair cover and a mask. We all wore heavy lead gowns to protect us from the cancer-causing X-rays.

The patient's head was behind a sterile green cloth. The only part of the patient visible to the theatre team was a patch of skin about the size of the palm of your hand on the right side of the groin.

The doctor would do a 'cut down' to carefully open the flesh until they could see the femoral artery pulsating. Then they would carefully dissect into this femoral artery, and slide in a cardiac catheter that was about 2 millimetres in diameter and a bit longer than a metre, ensuring the patient's high-pressure arterial blood did *not* squirt all over the operating theatre. The next step was to slide the catheter up the femoral artery into the curved arch of the aorta, push it through the U-turn, get it to the very beginning of the aorta and then manoeuvre it sideways to place the tip of the catheter into one of the two main coronary arteries. Yes, this was skilled and delicate work.

Then they injected a radio-opaque dye into the blood to see the blood flowing in real time. The image was captured by the X-ray machines and relayed to both the live viewing screen and the film camera.

I sat in a little anteroom next to the theatre recording various pressure measurements as the catheter was manoeuvred around the patient's cardiovascular system. I would also measure the patient's cardiac output. A healthy adult should be able to push about 5 litres of blood out of the heart every minute.

I liked answering the student nurses' questions. About twice each week, small groups would come through the theatre as part of their professional training. Initially, they would just stand silently off to one side, not able to see much and with nobody paying them any attention, and then they would leave. Unfortunately, the theatre staff were too busy to also do a bit of teaching.

I had little gaps in my time during each procedure, so I would use that time to explain what we were doing and why. If the nurses hadn't studied physiology yet, I would explain the electro-fluid mechanics of the heart, the cardiovascular system and anything else relevant to each particular patient.

My little chats had to be interesting, accurate and entertaining, and it reminded me of my teaching role in New Guinea. The nurses would ask deceptively simple questions, and when I couldn't explain the answer clearly, it forced me to see the gaps in my knowledge and I would have to go back to the books myself.

I think this is where the start of all of my science communication really began. Later, this would be the central theme of my working life.

I also ran the unit's film-processing machine. Each day, the theatre would generate several hundred-foot rolls of black-and-white film showing the blood flow related to the patients' heart conditions.

So what were we looking for? Our day-to-day role was checking out any blockages in the arteries that fed the muscles of the heart. But occasionally we would check out other cardiac problems, such as a hole between some of the chambers in the heart, or abnormal 'plumbing' of the blood vessels around the heart. (These were usually from congenital heart problems, meaning that they were present at birth.)

I gradually took over the film processing from the other scientific officer. It was usually straightforward: put film into machine, hit the start button, remove the processed dry film after an hour or so, deliver to the doctor. There was the slight complication that the film had to be loaded into the processing machine in a darkroom with the lights out, but that was easy to learn.

Occasionally, the drying phase of the processing machine would fail. (Intermittent problems in technology are the very devil to fix, because the technicians often can't find what needs to be fixed.) Sometimes time was critical, and the diagnosis couldn't wait for the half-day that it took the film to dry out naturally. So I brought the drying unit in from my own now unused home film lab into the hospital, a box about half a metre by half a metre by 1.5-metres high. It would get used maybe once a month, whenever the other one failed. I was happy to have my filmmaking experience come in handy!

One morning, while I was driving to work, the lead news on the radio was about an urgent ambulance transfer of a baby to a Sydney hospital for a possible 'hole in the heart'. You guessed it, that very baby turned up at my hospital to be my patient!

Just for background, *all* babies are born with a hole in the heart. It's between the right and left atria, and it exists so the blood flow in the heart bypasses the lungs entirely while the baby is still inside its mother. That means babies don't need to use their lungs to breathe until they're born. So the hole is essential for the unborn baby to be able to grow, but once the baby is born, the hole is a liability. So almost always, the hole shuts naturally within twelve to twenty-four hours and is permanently 'fused' shut by about three weeks. But sometimes, things go wrong.

On average, we would have a baby as a patient about once each month, but it never usually made the news. The entire team had heard about the baby coming in for assessment on the radio news, and we all guessed that it was just a slow news day. For us this was no big deal, and we treated the baby as we treated all other babies. Happily, the baby's surgery went well.

When I got home, one of my neighbours had obviously heard the news too and asked if I had taken any part in the care of this particular baby. I said I had, mentioning that in addition to my regular work I'd placed a warm blanket on the baby, and they were ever so impressed. Unexpectedly, a warm glow flooded through my body, and I realised then that I was doing good stuff.

On top of the regular five-day working week, I was also on call 24 hours a day every alternate week in case patients with a sudden cardiac 'event' needed an immediate diagnosis. On one occasion my pager went off because newly born Siamese twins were in distress and desperately sick, and needed an immediate diagnosis and urgent surgery. We started the procedure before midnight and finished only after sunrise.

Siamese or conjoined twins are identical twins, who get physically joined shortly after conception. They are named after Chang and Eng Bunker, who lived in Siam (now Thailand) in the mid-1800s. Chang and Eng were joined along the side of the chest and shared one liver. With today's technology, they could be easily separated and live normal separate lives (though not in the 1800s). However, Siamese twins who share too many internal organs can't survive separation. Tragically, our little Siamese babies fitted into this category and actually died during the procedure.

Very quickly afterwards, the whole theatre emptied, and I was left alone there with the twins. I guess everybody who had been involved for the last eight hours was heartbroken and needed a little bit of time out – plus, of course, the parents had to be informed. We had been going flat out, in a state of high tension for eight hours without a single break, but I felt I needed to keep a little vigil with the twins after they died. I felt sad, but I was also irrationally hopeful that I was doing good by being there with the little babies. Gradually, the staff trickled back and regular hospital procedures kicked in.

I got breakfast from the cafeteria and hung around for an hour staring blankly at the blue ocean, watching the sunrise. And then I started a regular day shift as though nothing had happened. Debriefing and supporting staff through traumatic experiences was not a big part of hospital routines back in those days!

The end

I loved my work. I especially loved helping the patients. At the beginning, while I was learning so many new things and then later while I was 'teaching' the student nurses, everything felt great. I got a kick out of being able to explain something well enough to make it clear for the other person. For me, the joy was in sharing information around so that we could all head down a common path to more understanding. Seeing enlightenment flash across faces was such a buzz.

But after a few years, I wanted more intellectual challenge.

Around then, microcomputers were slowly entering the Australian market. This was so early in the history of personal computers that the Microsoft company had just been founded. Basically nobody had a personal computer. In my own time, after work, I undertook two non-degree courses at the University of New South Wales in computer science and electrical engineering, which were basically on how to program and use microcomputers. A microcomputer, the forerunner to the home computer, was a box about the size of a book stuffed full of electronics, which could do simple jobs only, like adjust the temperature. This exposure to brand-new computer technology inspired me to have Big Ideas on how to improve the daily routines of the operating theatre.

One of these ideas was to send the information I could see on my screen in the little antechamber, to a monitor inside the actual operating theatre, in real time, so the doctors could see it too. I further suggested that having monitors to show the blood oxygen measurements in real time would minimise mistakes being made when these were only spoken out loud to the operating cardiologists. Video recording technology was also finally getting close to being as sharp as film, so I tried to organise an electronics company to demonstrate some of their high-end video gear to the staff.

But the boss's position was that everything was working just fine. My changes might work, but they also had the potential to go wrong and make things worse. And there is that old saying in engineering, 'If it ain't broke, don't fix it.'

I loved my co-workers and being part of Team Making the World a Better Place, and I loved my patients – but mentally, I was getting bored. It was time for a change, but to what?

That turned out to be another question whose answer came out of left field. As usual, it was a gift from the random noise of the universe …

1978–80

Me and eye

By late 1977, I felt like I had learned about as much as I could as a scientific officer in the cardiology catheter laboratory. It had been a great ride over a few fun years. I had picked up an idea of how the human heart functioned, worked with a lovely bunch of people and learned about the many different things that hospitals do. But I wanted something new.

Luckily, around then my parents had invited me to go on holidays with them to New Caledonia. (Prior to this, travelling with my parents was very low in my holiday priorities – I foolishly had usually said 'no' to their invitations to travel as part of the family on previous occasions.) We got on the plane at Sydney airport and hopped to New Caledonia, and then to the tiny island of Ouvéa.

There was nothing much to do except swim, eat and read. The swimming and eating were magnificent, but with practically everything available written in French, which sadly

I couldn't read, I started on the several-kilogram bulk of the weekend edition of the *Sydney Morning Herald* that we had carried all the way with us. I was so desperate for something to read that I even got through all the ads that made up most of its enormous heft. (Advertising in newspapers used to be a very big thing before the World Wide Web came into being in late 1990.)

Something popped out of the ads and grabbed my attention – a brand-new academic course at the University of New South Wales called a Master's Degree in Biomedical Engineering. I could immediately see the appeal. I liked the little I knew about the human body, and I loved making stuff with my hands. So combining the biomedical and the engineering parts of my life seemed a natural fit.

Once again, my life was unexpectedly heading down another track!

Hello again to full-time study

Back in Sydney, I immediately applied for the course, and in early 1978 I resigned from the hospital and started back at uni. I was still living rent-free in the squat, and back then there were no university fees (thank you, taxpayers of Australia!), so with the money I had saved in the bank I could live very frugally.

My three-year plan was to attend lectures full-time in 1978, do a mix of lectures and preparation for my thesis in 1979, and in 1980 finish building the engineering project (whatever it turned out to be) and write my thesis. I had no ideas at the start for a thesis topic, but I figured something would turn up.

> ## University degrees: an overview
>
> A master's degree is one above a bachelor's degree, but one down from a doctor of philosophy (PhD).
>
> Confusingly, a doctor of philosophy usually has never studied philosophy. You can do your PhD in a whole bunch of different fields (chemistry, architecture, or Mongolian literature) and gain a 'doctorate' and get the title of 'Doctor' without studying any philosophy (or medicine!) at all!

You can get a master's degrees in two ways: by coursework or by research.

In the coursework version you spend a few years going to lectures and tutorials, and if you pass all your assessments and exams you receive your master's degree. It's like a pumped-up bachelor's degree.

The research version is different, in that you also spend time on your own research project and submit an original thesis at the end. A thesis is usually based on reading stuff that other people have done and then taking your field of knowledge a bit further. Ultimately, you need to generate a whole new 'body of knowledge'! In an engineering thesis, you often also have to build a device that did not exist before, or at least an improved version of something existing.

My degree was a typical hybrid of introductory lectures, combined with an engineering project for me to build.

Because this degree was in biomedical engineering, there were two more choices about the engineering project – whether to build a therapeutic or a diagnostic machine. Therapeutic machines do something to make people better (like stimulate broken bones so they heal faster). Diagnostic machines diagnose diseases and medical issues (like a machine to detect broken bones).

I ended up building a diagnostic machine.

I spent all of my first year attending background lectures in statistics (because you have to be able to analyse data), pathology (the study of diseases) and other medical subjects, plus a compulsory arts course to round us science types out (hello, film appreciation!).

I *loved* film appreciation. The idea that just watching a movie and then talking about it in class was study seemed pretty attractive to me. And the grading for this course was wild! Everyone's name was put into one of two hats. One hat was for a pass or credit and the other for a distinction or high distinction! When I asked the lecturer to put my name into the latter hat, they said, sure, no worries. They didn't care at all. I've got no idea why everyone didn't ask for the best possible mark. My high distinction, which was literally pulled out of the hat, later turned out to be important for me getting into medicine.

Being a good student (for the first time)

Physiology, the study of the human body and how its various systems work, was huge both in content and course load, about sixteen hours each week. It also blew my mind!

I had no idea there was the equivalent of an entire universe inside the body. Every single organ and system is more complex, in its own way, than the Big Bang. Under your skin, you are a dynamic maelstrom of chemical, physical and electrical activity. That's incredible!

But the kidneys nearly killed me with their fiendish complexity. Physiology courses usually start with the kidney, to show the students how much they *don't* know! Students emerge after months of studying the kidney, blinking in the sunlight, realising that they still know barely anything about it.

Kidney 101

The kidneys help control blood pressure, help moderate the body's acid-alkali balance, make hormones, generate urine, get rid of some wastes, get calcium into your bones and so much more.

Just for starters, each day your kidneys filter about 150 litres of blood. In doing so they pull around 500 grams of sodium out, and then they put practically all of this back in except for about 3 grams, which ends up in the urine. All this work takes a huge amount of metabolic energy, and some very complicated kidney subsystems.

Why do the kidneys have to go to all this trouble? Well, because evolution made a

> mistake, and we humans are basically 'fish gone wrong'. Our kidneys evolved for life in salt water, and to survive on land, evolution added on an extra processing unit.
>
> Before multicellular life came onto the land, it spent nearly a billion years in the oceans. As a result, many of our organs and systems have 'heritage' features that came from that past. Evolution doesn't have to be perfect, just good enough, and good enough means having babies who survive and go on to have more babies. As life evolves, sometimes systems get a patch-up rather than a full rebuild!
>
> And that's what happened with our kidneys, which originally evolved for life that lived bathed in salt water, not on land.

This time around at uni, I was a good student who tried to focus! I did lots and lots of study sitting in my purple bean bag on the upstairs verandah of my Glebe squat, where I could still watch the daily goings-on in the street below and not feel too isolated from life. I was spending about sixty hours each week on my master's degree, including on-campus activities and home study. At the end of the first year, my science marks were mostly distinctions and high distinctions, and none of them were pulled out of a hat! I got them the old-fashioned way.

And then, out of the blue, at the beginning of the second year I was invited to an interview with Professor Fred Hollows, who was linked to the University of New South Wales – where I was studying.

At the time, Fred was a professor of ophthalmology (an eye doctor) who had taken on the extra personal goal of delivering eye care to Indigenous Australians. He was spending a lot of time in the Australian outback treating Indigenous people for eye diseases such as trachoma, an infectious eye disease, to try to save their precious eyesight. He worked his whole life in remote conditions in Australia and overseas, treating eye disease and teaching and training others. His plan was to make simple sight restoration procedures available to many more people around the world.

He once told me something that has stuck in my mind. Indigenous Australians have some of the sharpest eyesight ever measured in humans, but they are also the most likely to be blind in their later years from entirely preventable and treatable medical conditions like trachoma.

The 'standard' best eyesight is 6/6. This means that you can see at 6 metres away what the average person can see at 6 metres away. (The measure 6/6 is the metric version. The imperial version is 20/20 – you can see at 20 feet away what the average person can see at 20 feet.) Occasionally, people with better eyesight have been measured at 6/5, or 6/4, or very rarely 6/3. But some Indigenous Australians have eyesight measured at 6/1.5! This means they can see at 6 metres away what the average person can see only at 1.5 metres. Even more astonishing, this sharp vision is not just in the young but also in the elderly.

Evolutionary biology has an explanation for this incredible long-distance vision. Water is scarce in the deserts of Australia, so it was essential to be able to recognise landscape features associated with waterholes. When a young person walked through an area with water, they learned and remembered the geography. Twenty or forty years later, if they were back there again, they needed the memory that those imperceptible bumps on the horizon meant water. And they also need the sharp vision (even as an older person) to still see those microscopic bumps.

Fred came straight to the point – which, I would learn, was pretty typical of Fred! He said that I had very good marks and wanted to know if I could build stuff.

I said, 'Sure, I can fix cars and do electronics. What kind of stuff do you want built?'

'I want a machine that picks up tiny electrical signals from the human retina – and by tiny, I mean millionths of a volt. It would be used to detect retinitis pigmentosa and other diseases.'

What an opportunity – and a challenge! Prior to this, I had no idea what I was going to build for my master's degree. Now, out of nowhere, I had an offer to build a machine for Fred that would meet the engineering requirements of my master's degree, and I could do the coursework in tandem.

'Sure,' I replied. 'Give me a year to find out what other people have done and another year to build it. I don't want to waste time reinventing the wheel.'

'Great,' said Fred. 'There's also a scholarship with money. We'll send you to North America where they do this stuff, and then you can come back and start building.'

What a bonus! The small scholarship would cover my microscopic living costs, with a bit left over for giggles and fun.

I spent the first two-thirds of 1979 doing my regular biomedical engineering coursework while starting my own reading on vision. Eyesight is an incredible sense! I studied how light comes into the eyeball to land on the retina, gets converted into electricity, and three-tenths of a second later gives you glorious 3-D wall-to-wall colour vision. I was reading about the anatomy, biochemistry and electroneurophysiology of the eye in textbooks and scientific and medical journal articles.

Today, to get access to a scientific paper, I simply go online, search for it and download the article. It takes a few minutes at most. But back in the 1970s, it could take days to get hold of an article you needed to read. I had to walk to the library, look up the name of the journal on microfiche (ultra-high-resolution photographic film), and if I was lucky the university library kept the journal on the shelf and I could photocopy the article. But sometimes that issue might be stored in the basement, and I had to wait a few hours for a librarian to get it.

Sometimes you even had to apply for an inter-library loan. The librarian faxed another library somewhere else in the world, the distant librarian found the article, photocopied it and then faxed it back. After a day or two, you could finally get hold of the article!

The internet is good for fast access to real knowledge – if you know where to look!

Electroretinography 101

Fred wanted me to build him an electroretinograph (ERG) machine. I knew absolutely nothing about ERGs before I began.

'Electro' means electricity, 'retina' is the layer covering three-quarters of the inside of the globe of the eyeball that turns light into electricity, and 'graph' means to write. So an electroretinograph (ERG) is a machine to pick up and record electrical signals from the human retina. The very first human ERG was made way back in 1877, but it took a while to understand what the signals actually meant.

Why would you want to measure electrical signals off the human eye? Well, one in seventy people carries the gene for the eye disease retinitis pigmentosa, and roughly one in every five thousand babies is born with the disease. The symptoms can start in childhood with loss of night vision and worsen with increasingly tunnel vision. You usually get diagnosed as a teenager. There are at least thirty different varieties of retinitis pigmentosa.

'Loss of night vision' might sound weird – after all, none of us can see at night, in total darkness. But imagine a group of teenagers walking home in the semi-darkness, with streetlights every 50 metres or so. Most of the group can see the parked cars directly under each streetlight *and* the parked cars in between the streetlights. But a teenager with retinitis pigmentosa can see only the cars in the pool of light under each streetlight. In between, they see nothing – just total blackness. If they walk into a parked car (that they can't see), they end up getting an injury that might unmask the underlying retinitis pigmentosa.

Retina 101

Most of us have two major types of cells in our eyes to turn light into electricity – rods and cones. They're both long and skinny like cylinders – the tips of the cones are slightly tapered, and the tips of the rods are not.

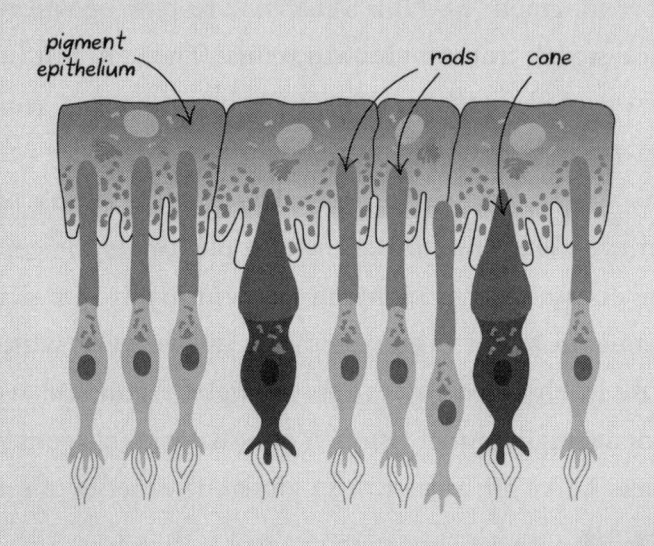

pigment epithelium — *rods* — *cone*

The rods come in one variety and there are about 92 million of them. They respond only to blue light, so they don't work in the daytime. But at night they start to work after a few minutes of darkness and reach full sensitivity after about forty minutes.

The cones come in three varieties and are sensitive to blue, green and red light. (I remember them as 'C for cone and C for colour'.) You have about 6 to 7 million cones and they function best in daylight, but they will work a little in the lower light levels around dawn and dusk.

You can do a simple experiment with light and colour. Sit outside during the shift from day to night, with no streetlights around, and pick out a red car, then keep an eye on it. Gradually, you should see the red colour change to a blue-grey shade. The red-sensitive cones stop working in the dark, and now your blue-sensitive rods are seeing the car.

People with retinitis pigmentosa lose the rods in their retinas, so they lose night vision.

The retina is smart: it doesn't just turn light into electricity, it also compresses and processes that electrical signal as a genuinely very sophisticated computer. This processed electrical signal is sent from the retina to the visual cortex at the back of the head. The visual cortex is made of two little lumps of brain tissue (one on the left side of the brain and one on the right), each about the size of a baby's clenched fist. The visual cortex finishes the job of 'seeing' by turning the incoming electricity from the retina into the

> glorious 3-D colour experience that we call vision. This extra processing is so complex that it can't happen instantaneously – it takes about three-tenths of a second.

Retinitis pigmentosa can be diagnosed by looking at the electrical signals given off by the rods and cones. But how do you measure something from the retina, which is at the back of the eyeball?

Well, we already measure electrical signals from the heart, which is deep inside the body. Millions of people have an electrocardiogram (ECG) each day. The inside of the human body is full of salty water, so it carries electricity really well from the heart to the skin. For an ECG you just need two wires sitting on the patient's skin, an 'active' that carries the actual signal and a 'neutral' that completes the circuit.

But getting an electrical signal off the retina? Patients would run a mile if you approached their eyeball with wires! Getting an ERG seemed to be a lot more complicated than getting an ECG. I had to fully understand the current state of the art in detecting electrical signals from the human retina, and then take that one step further.

Visiting the laboratories of world leaders in this field would be a good start. And thanks to the scholarship via Fred Hollows, I could afford to!

So towards the end of 1979, I flew to the US and visited laboratories in Seattle, Vancouver and New York. I had a month working at the Columbia University College of Physicians and Surgeons on 168th Street, at the northern end of Manhattan.

In New York, I was lucky enough to work with somebody who was doing groundbreaking research in electroretinography. I became his student and always-available test dummy!

We took measurements from my retina every single working day and managed to pick up signals from the blue-sensitive cones, something that had rarely been done well. This involved 'floating' an electrode on my cornea, the delicate front layer of my eye, for extended periods. My cornea did get damaged by the electrode, but it recovered within a day because the damage was superficial.

Each day we tested alternate eyes, and at the end of each session my boss gave me antibiotic eyedrops and taped that eye shut overnight to protect it. The next day, he would test my other eyeball.

I wore an eyepatch over the taped eye every workday. I'd walk into the lab in the morning wearing a patch on one eye and leave in the afternoon with a patch on the other eye to return to my accommodation in the Nurses' Home. One night, a nurse administrator called me over and said accusingly, 'Look, you're not fooling anybody.'

Bewildered, I said, 'What do you mean?'

'Well, you're clearly wearing that eye patch to try to get some attention,' she said. 'But you keep forgetting to wear it on the same eye, so we all know what you're up to.'

I had to laugh. Once I explained my research to her, everything was cool.

I spent a wonderful month in New York City and loved it to pieces. I still do. New York City is a strange place. It's not like other parts of America, and it's not even like the rest of New York state. It just is what it is!

The nurses showed me how to keep my yoghurt cool in winter by putting it on the outside windowsill. (New York in winter is freezing!) They took me to the live theatre, both Broadway and off-Broadway. We walked all over Manhattan seeing the sights, and they introduced me to all the different styles of food available.

But New York also had a dark side. Around Christmas, one of the scientists (who was also a Catholic nun) casually announced at morning coffee that somebody had dumped a body at the door of the nunnery where she lived in the Bronx. The boss, who lived in Riverside, a much classier area on Long Island, chimed in with, 'We had a murder in our street last night as well.' Then they switched to discussing the research. Apparently, two murders on one night in New York was nothing much to talk about!

Panic stations!

I had promised Fred I could build a world-class electroretinograph in a single year! And when I said yes to him, I really thought I could.

But once I was back in Sydney having visited three state-of-the-art, fully functioning electroretinography laboratories, and had seen what large teams had independently put together over a period of several years, I changed my mind. It was such a specialised field that nobody sold ERG machines commercially. If you wanted one, you had to build it yourself!

A Periodic Tale

I had bitten off more than I could chew and couldn't do it on my own in just one year. I realised I needed full-time technical help from somebody with very rare skills. The bare minimum requirement was for my colleague to have every relevant mechanical, electrical, electronic and computer skill known to humanity in 1980. And they also needed a brain that could quickly pick up all the necessary visual electro-neurophysiology to understand the overall goal. That's some job description for a random somebody!

Luckily, I already knew Jackie Joy from my hippie days. I used my grant to employ him, beginning in early 1980.

Jackie was brilliant and strange. If Armageddon came and you wanted to rebuild modern society from the Stone Age, Jackie could save your bacon – all the knowledge needed was in his head. He taught me that the two basic tools needed to make anything were a hammer (to make stuff change shape, or weld metals together) and a file (to remove stuff you didn't want).

Jackie was a wild card. He loved hotdogs, salami and science. He also loved numerology, and he changed the spelling of his name to bring him more luck. At one stage he lived in the roof space of a mutual friend's shop and sent down his bodily wastes via a little pulley system every morning. Why didn't he just come down the ladder and use the toilet? You didn't know Jackie! Rather than a short shower every day, he would have one long shower every week. He always wore white overalls and had rubbed silicon rubber into them and let it cure, so he didn't need to wash them ever again – a simple wipe with a wet cloth was enough. He ate all his food from the same bowl, mixing cereal, milk and orange juice, and claimed that made sense because it would soon be all mixed together in his stomach.

Employing Jackie solved one problem, but there was another big problem – I didn't have an office. My repeated requests for an office-cum-workshop were considered very low priority by the university, because I was just a student.

My supervisor was in the School of Medicine, so that was where I was based. So one Monday morning I came in to work extra early. I dragged a spare desk and chair into one of the two elevators and sat at it all day, going up and down hundreds of times while theatrically doing my work, but also being appropriately chatty with everybody who came into my new 'office'.

Word got around very quickly, and by that afternoon the head of school told me that, by an amazing coincidence, a huge room that filled half the top floor had very recently fallen vacant. The head needed to keep that enormous space occupied by a 'placeholder' for the rest of the calendar year, because major renovations would happen in early 1981. Was I interested?

Oh yes indeedy! The space was huge, with fantastic views of Randwick racecourse, the Sydney CBD, the Harbour Bridge and the Opera House. It was probably the best office I've ever had in my life.

Jackie settled into the university routine with me very quickly. He quickly found the machine shops, and used university lathes and milling machines to make a few specific parts that simply could not be bought. He designed and built multiple complicated components and was intelligent and compassionate enough to come up with a whole bunch of user-experience design features that I hadn't thought of – like painting everything baby blue, just like IBM (see 'School years' chapter for the full story on that).

We worked day and night, seven days a week. We met up with Fred regularly to keep him up to date with our progress. Jackie and Fred got on like a house on fire. Being a bit of a free spirit, Jackie went exploring through the nearby Prince of Wales Hospital and found an unused room. Without telling either Fred or me, he started sleeping there. (At least he was now using the bathroom, like a 'regular' person!) He got busted after a month, but there were no hard feelings from Fred, who thought it was funny.

It took us about nine months to finally build the ERG machine. It weighed about 100 kilograms, was about a metre high and wide and 1.5 metres long, and it could be easily rolled around on its four steerable wheels. It was our big baby, taking the usual nine months of gestation!

Then came the next step: testing it on real patients. Happily, after a few 'shakedown' glitches, the machine was working perfectly thanks to our careful, precise work.

I had been writing up my thesis while we were building the machine, handing the chapters in to my supervisor as we went along. And then something random happened that changed everything!

Our machine was calibrated specifically for patients who had retinitis pigmentosa. But out of the blue, a motorbike rider came our way who had been unlucky enough to have a stone thrown up by a passing car land with terrible precision on his cornea. The front of his eyeball was an opaque mess and he couldn't see with that eye.

He needed a new cornea, but nobody knew if his retina had been damaged in the accident. They didn't want to waste a cornea if his retina didn't work, as cornea donations were precious and rare. So his eye surgeon came to ask if we could test if his retina was in good nick.

Luckily, we had greatly over-engineered the capabilities of the machine. It had the brightest available xenon arc flashtubes then available. In all our previous work with regular retinitis pigmentosa patients, we had the light output massively throttled down, and at this setting we didn't detect any electrical signals off our motorbike rider's retina. (That made perfect sense because the front of his eyeball was opaque, not transparent like a healthy cornea.)

Organ donation

Dear reader: please consider offering your body parts after you have finished with them to be used again to help those who are still alive. The technical term is organ donation. In Australia you can register to be an organ donor online through various state government sites. And tell your friends and family that you want to donate your organs as well, if you can.

Then we cranked the light output up to maximum, something we'd never done before. Jackpot! We could see a definite signal from his retina showing that it was functioning just fine and in all three colour bands of red, green and blue. He was a good candidate for a corneal transplant – and pretty soon, he would be able to see again.

He was happy. I was happy. Jackie was happy. What the heck, we were all happy! And then he said the prophetic words: 'So when are you going to do the operation?'

I didn't quite understand and said, 'What do you mean?'

He said, 'When are you going to do the operation? To give me a cornea.'

'Oh, I can't do the operation. I'm not a medical doctor.'

Motorbike boy frowned at me. 'Why not?'

And I asked myself, 'Why not?'

It literally had never seriously crossed my mind. I'd had the opportunity to study medicine when I left high school, but I dismissed it with 'I don't like blood'. That wasn't really a genuine excuse and, having worked in hospitals, I'd grown past that sort of worry.

So why not? I couldn't think of any reason to not be a medical doctor. Sure, I had enjoyed being a hippie, and a filmmaker, and a roadie, and a taxi driver, and a hospital scientific officer as well as an academic and a physicist. But I also enjoyed learning to understand the workings of the human body, and I loved diagnosing problems with it. Becoming a doctor was so obviously the next step.

At the time, I was well on the home run to finishing my thesis. Word had got around about my work, and I was lucky enough to have not one but three separate offers of a PhD scholarship with different supervisors to continue with my work in visual electroneurophysiology. That was incredibly flattering and totally beyond my expectations, but the die was now cast. I had committed in my heart to getting into medicine.

My next-door neighbour in the squats told me I was foolish. After all, I was thirty-two, and I would be forty-two before I knew enough to be a medical doctor. (Yes, the body of knowledge needed to be both competent and safe as a medical doctor is so enormous that it takes about ten years to load it all into your brain.) My dad also wasted no words telling me that the chance of me ever sticking with anything was next to zero!

I slept on it. But when I woke, the answer was clear. I told my neighbour, 'In ten years I'll be forty-two whether I do medicine or not, so I'll do it.'

In New South Wales at that time, there were only three places I could study medicine.

The University of New South Wales rejected me because of some complicated university law about not counting the grades I got during my master's degree, because I had a supervisor who could have helped me get those grades (which seemed pretty irrational to me).

I also applied to the University of Newcastle because I had heard they had a very different and creative way of teaching medicine. The course sounded magnificent, so I sat their special exams. I failed the test. In their feedback they said that I lacked creativity, which is funny to think of now that I'm a published author many times over!

So I also applied to the University of Sydney. They said, 'If you've got a distinction average, no worries. You're in.' Of course, there was the minor formality of filling out all the paperwork.

The official offers came out a few months later. I went to the newsagency on Oxford Street at 1 am, which was one of the first places in Sydney that the newspapers were delivered. Sure enough, there was my name in the newspaper saying that I had been accepted into the University of Sydney to do a combined bachelor's degree in medicine and surgery. I didn't go to sleep at all that night. I was just absolutely delighted. In fact, I was so happy that when I went to work at the university later that morning, I told a fellow student my great news and then piggybacked her all the way to my office on the seventh floor and back down again via the stairs in sheer excitement!

That offer to study medicine started me on a pathway to becoming a medical doctor, rather than a 'real' PhD doctor. And it led to the next big chunk of study in my life as a medical student.

Part 3

The career-hopping, family-focused years

I chose to study medicine, even though my father was dead set against me heading off in a new direction. He thought that every time I changed careers, I was throwing away all the intellectual involvement that I had invested in the most recent role. I guess he was exasperated that after he had finally got me into a decent job as a scientific officer, I had tossed it in to become a student again. And then after successfully completing my master's degree, I was again throwing everything away and starting from entry level in a new career. I don't know how he would have coped watching me continue to float in and out of different roles over the coming decades.

But from my point of view, there was a natural progression in my study. I was following a thread from the 'pure' sciences into the biological sciences, so becoming a medical doctor seemed an obvious next step to me.

I really thought that I would simply do my medical studies, become a medical doctor and then try to find my niche – general practice, gastroenterology, neurology, surgery or something else – and stay there until I retired. Hah! It was not to be.

For a start, while I was a first-year medical student, I got myself into radio – and never left. Surprisingly, that's been the longest continuous job of my life. It all began with me reaching for the stars – or more literally, applying to become an astronaut for NASA on the space shuttle. These radio programs began as three-minute prerecorded stories in Sydney youth station 2JJJ under the title 'Great Moments in Science'. Over several years these expanded into live science question-and-answer sessions. Now, more than four decades later, I am doing about ten live radio shows every week. So that was two jobs – doctor and radio.

Those radio stories evolved into written stories for my first book, *Great Moments in Science*, which birthed my third parallel career as an author. This book is number forty-eight!

To publicise my first book, I cold-called the daytime TV program *The Midday Show* and asked if I could come on as a guest. That blatant self-promotion evolved into a fourth parallel career as a TV reporter. Yep, still doing that too!

So in fact, over the next four decades I returned to the habit of my hippie years. I worked in the kids' hospital as a doctor for about five years, but also had a bunch of other parallel careers, including radio and TV presenter, journalist and author, four-wheel-drive tester in the Australian outback for a few decades, public-speaking, academic and more.

But let's start with …

1980s

Love, loss and launches

In late 1980, master's degree in hand, I decided to abandon a career in biomedical engineering and strike out on a new path to become a medical doctor. I told my parents that I had applied to several medical schools. Out of the blue I received a letter from my father, mailed to my squat in Glebe. It was typed laboriously on his beloved typewriter (which I still have, and also love to pieces – he liked the twist that it came from a former German machine-gun factory that now made high-quality typewriters).

It started off with:

> Mummy told me that you have lodged an application for medicine faculty at Newcastle. I will tell you frankly: it is rather a bitter disappointment for me.

He went on to list in harsh detail how I had wasted a lot of time in my various careers:

> And what are you? Unsuccessful 'filmmaker', unsuccessful New Guinea tutor, unsuccessful taxi driver, unsuccessful heart researcher and so on. Now longing for the glorious day when you become (if …!) a doctor.
>
>   A tremendous CAREER, isn't it? Your tremendous CAREER achieved <u>22 years after your Leaving Certificate</u> and <u>15 years behind your schoolfellows who long ago became doctors</u>. And 25 years before your retiring age. It is a really tremendous CAREER!

Ouch! He was factually correct, and it stung. In his eyes I was years 'behind' my fellow school students, who now had established careers, families and owned their own homes. By contrast, here I was, with no job or income, living in a squat and planning on having another five years without income! I had microscopic savings when I started medicine. I was turning thirty-three years old and had just thirty-three cents in the bank! Poetically pathetic!

He also pointed out that if I were to stick with studying medicine, by the time I was able to become a practising doctor it'd be only twenty-five years before retirement. (He didn't foresee the extension of the usual working life that was to come, nor my desire to work forever!)

But here's the rub. My father had done similar things, jumping from lawyer to journalist to Hollywood scriptwriter before ending up in Wollongong on the other side of the equator from where he was born, and settling into a comfortable niche.

> For me one thing is sure: I won't survive the day of your glorious CAREER. I am feeling miserable and deteriorating. I am not sure if I even will be able (my blood pressure and loss of strengths) to say you goodbye when you triumphantly will move to Newcastle towards your new GLORIOUS CAREER.

He was right to be worried about his health. Within a year of writing this letter, he was dead.

At the time, I didn't know how to respond. So I stuck the letter back in its envelope and forgot about it. In retrospect, I wish I could have talked to him there and then. But I don't know if I could have done anything to change his mind and help him understand my dreams. Instead, I stuffed my feelings in the same envelope as the letter and headed off to med school.

Med school and Mary

In 1981, at the age of almost thirty-three, I started in first-year med at the University of Sydney. There were about 250 of us enrolled in that year.

I was different to most of the group. I was obviously not straight out of school, and I had a fair bit of life experience behind me. My new fellow students were not the kind of people I previously rubbed shoulders with. About two-thirds of them had come straight to uni from private schools. I had never really mingled with groups of mostly wealthy people before. There was one student who was sixteen years old in first-year medicine, so yes, I was double her age!

I made friends easily, because I didn't fit into any of the 'in' or 'out' groups. Basically, I was still the same old hippie, with an open heart and an open door to my squat.

During one very atypical week, a temporarily homeless med student shifted into the spare room in my squat on the weekend. Over the next seven days, for the first time ever in my life, I had a different lover stay over every night. In fact, so many different people came through, that my new house guest packed up and left in shock the very next weekend. I was as surprised as he was by the week that had unfolded. But I was even more surprised that he would be so unsettled by my apparently loose morals, that he would rather be homeless than live with me!

Right from the beginning of medicine, I always sat at the front of the class so I could see the blackboard. I am super short-sighted and can't see anything much without my glasses on, so I always plonked myself down in the front row. In classic form, there were 'good' kids at the front and a group of 'bad' students at the back who sailed paper aeroplanes down to the front of the lecture theatres. This went on for not just days but weeks.

I was hit – painlessly – on the back of my head many times. However, I was genuinely worried that if I were to turn my head to the side, a paper plane could come in the side of my glasses and hit one of my eyeballs directly – which could permanently scar that cornea. I already had bad eyesight – I didn't want to risk it getting any worse. So after about a month of being hit by paper planes, I asked the chemistry lecturer if I could have two minutes at the start of class to discuss my issue. She was fine with it.

I went up to the blackboard and, using chalk, drew a picture of my head in profile and went through the stats. I had been hit on the back of the head about twenty times. My eyeball was about one-fiftieth of the cross-sectional area of my head. So, after another thirty paper plane impacts, the chances were very high that my eyeball would be hit by the pointy end of a fast-moving paper plane. I asked my fellow students to stop throwing the paper planes so that I could exercise the option of turning my head from side to side, rather than fixedly staring straight ahead. I figured that if I gave them the logic of my request, they would appreciate that I was more than some party pooper.

Unbeknown to me, Mary, my future permanent beloved, was sitting in that class too. She went home and told her mother that a really tall student in short shorts and odd socks had been talking to the whole class about paper planes. From this somewhat obscure reference point, her mum decided that Mary was in love with me, and never budged from that proclamation. My mother-in-law-to-be was years ahead of Mary and me finally hooking up and telling each other that we were in love!

But Mary stood out for me too. Like most of the first-year students, she was straight out of high school. She was tall and had a face that shone like a glowing sun. When I first saw her in first-year medicine, I was washed by a feeling that had never happened to me before. No other female person that I had ever met had a face that shone like Mary's. The random universe smiled on us when we met each other at university.

> **The neurochemistry of falling in love**
>
> I hear people making extravagant declarations about brain chemicals and their effects all the time. It sends me a little crazy.
>
> People talk about a 'serotonin rush' making them feel great, or the 'love hormone' dopamine (or oxytocin) knocking them to their knees, or the 'runner's high' of endorphins giving them so much extra energy. (Anyway, 'runner's high' comes at least in part from natural cannabinoids too!)
>
> No, it's much more complicated than that. Which is really what you'd expect! We are slowly learning how the brain uses chemicals, electricity, magnetic fields and more, but we are still at the beginning of a very long journey into the mystery of the human mind.
>
> Around 1900, scientists suspected that the brain ran on electricity (partly true).

Electricity runs in a closed circuit, so when neuroanatomists later discovered that nerve cells in the brain didn't physically touch each other but had consistent tiny little gaps between them of about 20 to 40 nanometres, they needed to think again. (A nanometre is a billionth of a metre, or roughly the size of a molecule of glucose, which has twenty-four atoms).

The first neurotransmitter detected was acetylcholine around the 1920s. (A neurotransmitter is as advertised – it transmits information from one nerve to the next.) One nerve releases billions of molecules of acetylcholine, that quickly drift across the tiny gap to land on the receptors on the second nerve. This process is then repeated. Back in the 1970s, everything going on in the human nervous system was supposedly explained by that one single neurotransmitter, acetylcholine – at least in the popular literature.

By the time I started medical school in the 1980s, there were a dozen or so known neurotransmitters, but today there are hundreds – and the number is climbing all the time. They include glutamate, nitric oxide, dopamine, serotonin, oxytocin, histamine, endorphins and even cannabinoids (which are very similar to cannabis).

Neurotransmitters can have more than one action depending on the circumstances. The first identified neurotransmitter, acetylcholine, can have very different effects depending on its concentration, the presence or absence of other neurotransmitters or drugs (such as antibiotics), your blood calcium levels on the day, and much more. So acetylcholine can work either as a stimulant or a depressant on your brain activity and it can also be short- or long-acting – and that's just a few of the possible variations. It gets even more complicated when you combine the myriads of neurotransmitter chemicals on the different types of brain cells.

In late 2023, the Brain Atlas project (formally called the Brain Initiative Cell Census Network, or BICCN) announced it had identified over 3300 different types of cells in the brain! So can the incredibly complex state of 'love' be purely due to the action of the single neurotransmitter, dopamine, on a single cell? The answer, in a heartbeat, is no! It's much more complicated.

Mary and I started out platonically as study buddies. But along the way, there were a fair few times when we would touch fingers or elbows as we leaned over the pages of a book. And both of us were unsure if it was accidental or suggesting something more interesting! By the time she was 21, we were lovers. To my surprise, in fourth year, I found myself surreptitiously holding hands with Mary beside patients' beds.

There was a window at the start when we were definitely off and on. I once wrote her a letter while I was away travelling in Japan to tell her that she was the most wonderful person in the solar system, and I meant it. But on returning to Sydney, I broke up with her and immediately went out with someone else. I don't know why. I guess I was phobic about commitment, and our relationship seemed to be heading too far down One Love Street.

But somehow, we always made up and started again.

We share common beliefs in fairness, generosity and loyalty, along with a firm love and devotion to each other. Luckily she is the kind of person who can accept me telling her she is as beautiful as a white whale and know deep down that I mean it as a compliment. Which I really do. I love whales!

Despite a shared essence, we go about things very differently. I send her crazy with my desire to look at every single possibility before I make any decision! Even something simple like parking the car takes me ten times longer than she is willing to wait; there have been plenty of occasions when she has run out of patience with me straightening the car perfectly and has had to get out. Probably we annoy each other every day, but one of us is always willing to say 'let's start again' or to make the other person laugh, and then the fight is gone.

My mother was never keen on us as a pair. She saw Mary as an outsider and a 'peasant' who threatened everything that she held as important in our mother–son relationship. That was a tension that I regret for both of them. Almost certainly my mother's fears were rooted in her terrible experiences in concentration camps, and the fact that she just didn't trust anybody outside our family.

Mary is the oldest child in a large Irish Catholic family. She has five brothers and infinite numbers of cousins and aunties and more. As an only child, I could never work out all her family relationships, or how I was connected to everyone else through her. (Even now I still occasionally refer to my hand-drawn family tree to work it out.) But despite my inability to grasp even the most basic aspect of family relationships, I found warmth and inclusion in Mary's extended family network.

I hadn't ever really had children as a part of my life before meeting her younger brothers. They were so important to her that I suddenly saw first-hand how healthy extended families can function to support each other and offer many positive relationships. I now have so much more empathy for people who don't have a helpful family network nearby.

But let's get back from Mary to medicine. The first three years were totally academic, and we didn't see living patients at all. Anatomy labs were run with vats of human tissue floating in formalin, ranging from whole bodies to individual parts like a hand. The skin and flesh were a weird grey colour from the formalin, and there was a very particular smell in the labs.

Some students found this confronting, but I didn't. I saw all this study as adding to my knowledge banks, and that's what I wanted.

I wasn't quite as dedicated a student as I had been in my master's degree, as I found myself distracted by all the other options opening up in my life. I usually wasn't worried about failing any subject, but I wasn't getting high distinctions anymore either.

I did make a strange self-diagnosis, though.

All my life, I knew I was slightly different from other people, especially when it came to recognising and putting together faces and names. Everybody else I knew could do it effortlessly, but somehow I could not. I'd previously just assumed this was because of my tendency to be a bit solitary and distractible.

It all came to a head one afternoon in first year when two fellow medical students came up and said g'day. One of them asked, 'Hey, Karl, what's my name?'

'Your name is John,' I said.

And then the other one said, 'Then what is my name?'

'Well,' I said slowly, 'your name is John too.' (Meanwhile, I was thinking to myself how odd it was for them both to have the same name.)

The second John said, 'How come you think we have the same name? Can't you tell us apart?'

I looked at them, and my first answer was, 'No.'

As I stared at them both some more in confusion, I gradually realised that they were both tall, young, Anglo males with red hair, but one had glasses and the other did not. And then as

I continued staring at them, suddenly, as if out of a fog, the features on their faces became more distinct. I said, 'Oh, my God, you're different!'

The second one laughed and said, 'Yes, I'm Andrew.'

The other one said, 'And I'm John, and you're always confusing us. I'm the one with glasses.'

'Thank you,' I said. 'I don't know what's going on, but thank you for telling me.'

As they kept talking and laughing, the more I looked at them, the easier it became to see that they did look different from each other. But at the start, there was no way I could pick them apart!

It was a few years later when I found out that this was a recognised medical condition, called 'prosopagnosia'. Prosopagnosia is also known as 'face blindness' – you're unable to distinguish faces from each other.

Many years later, the producer on my science television show *Sleek Geeks* tricked me with my prosopagnosia. When I was attempting to break the Guinness World Record for signing the largest number of books in a single day, he got a random person to return to me over and over again in the book-signing queue – it took about ten times before I clicked and realised it was the same person each time!

To me, at first sight most people's faces look completely unremarkable, as similar as one brick in a wall is to another. But I have learned a few tricks to tell people apart. I write down someone's name when introduced with a note about any specific features they have, like curly hair or a beard, where they sit in the office or how they walk or maybe swing their arms. Then the next time we meet, I can check my little seat map to see who I am talking to. And I always say 'Hi, John' or 'Hi, Jane' with a bit of a quizzical lilt, just to make sure I'm on the right track. (Because maybe somebody else is sitting in their chair!)

I realised how bad my prosopagnosia was on a three-day comedy course (I highly recommend that everybody do this, by the way). On the first day, there was a guy sitting across the room whom I thought I maybe knew. On the second day I was thinking, 'Yeah, I'm pretty sure I know him.' And on the third day I went up to him and said, 'Hi, my name is Karl. I think we might have met.' And he said, 'Yes, Karl. We've just spent the last six months working together five days a week shooting *Sleek Geeks*. I'm your producer.' And as he spoke, it was as though a mist cleared, and details of his face appeared and suddenly I could see and recognise his features. Having worked with me, he knew I didn't recognise faces very well – and laughed it off.

The way I deal with not recognising people nowadays is just to be upfront: 'Hi, I've got prosopagnosia and I really can't recognise faces. Have we met?' And sometimes they say, 'Yes, we had a cup of coffee last week.' I can recognise my family, but every now and then I'll get their names wrong! And now I have the habit of secretly whispering to Mary, whenever I see people waving at us, to check who we are about to say hello to!

I'd just always thought I was bad with faces. But other people thought I was rude when I couldn't recognise them no matter how many times we had been introduced before.

So it was an absolute relief in first-year medicine to realise that I had a real condition to explain why I couldn't remember people by their faces.

In our fourth and fifth years of study we did separate blocks of clinical training in hospitals.

Then, if we passed our exams at the end of fifth year, we would suddenly become interns, working as junior doctors in the hospital system and wishing fruitlessly that we had more supervision!

Patients at the Concord Repatriation General Hospital where I was a medical student, and later an intern, were incredibly generous to students. 'Give it another go,' they would say when you missed the first attempt to put a canula in their vein! Once I walked past a patient, and a young intern had fallen asleep with their head on his bed, obviously a wreck from the long hours of study and ward work. I went to wake them, but the patient stopped me and said, 'Don't. They need to rest.'

While I got on well with most of the students, the relationship with the doctors who taught us in the clinical years in hospital was a bit more hit and miss. I wasn't bothered by the hospital hierarchy that placed medical students at the bottom of the pile, expected to watch silently and only speak when spoken to. I asked questions all the time. (I really believe passionately that asking questions is the best way to learn and that there is no such thing as a dumb question – not having a specific type of knowledge does not make you stupid.) But a tiny minority of the doctors found my questions put them on the spot, and they didn't like the fact that they couldn't keep me in line with mockery or derision. In my opinion, if you can't explain something, then you don't understand it. I reckon it is better to say you don't know than try to bluff it with bravado!

For some reason, the surgeons tended to like me, which is funny because hospital mythology has the surgeons being brusque and off-hand. But generally I got on famously with them. I loved all the tools they used to operate with, and the straightforward approach they had when interacting with people.

The thing I most remember about med school – apart from meeting Mary, the person I wanted to stay with for the rest of my life – was that medical doctors had to carry the most enormous body of knowledge in their heads! It was different to working in research or other science fields. In an emergency, doctors didn't have the luxury of looking things up – they had to know the answer instantly. So it was my goal in medical school to load my brain up fully enough to be able to give encyclopaedic answers to any medical question I was asked.

My parents' deaths

My father died when I was in first-year medicine. His death affected me in totally unexpected ways. For one thing, overnight all my teeth came loose!

My father had been writing letters warning me that his health was failing. He even told me that he was not sure if he'd get to say goodbye to me.

He was right about his health going downhill. Shortly after I started studying medicine in 1981, he had a stroke. I dropped everything and went to Wollongong. I missed out on a whole bunch of lectures, but I wanted to be at the hospital with him.

I kept visiting on weekends, but he didn't really get much better. One weekend, we thought he was improving, and then suddenly he started spiking massive fevers up to 41 degrees Celsius that the hospital couldn't control. After each fever he was significantly worse. He died in hospital within a few weeks of having his stroke.

It was a relatively fast way for him to die, and it saved him suffering a more painful death. We later found out from his autopsy report that he had cancer in his pancreas that had already spread to his liver. Pancreatic cancer tends to be nasty; it presents late and is a painful way to die. So in a way it was a comfort that his death was quick, not lingering. Fred Hollows once told me that everybody has to die, but what matters is that they have a good death. My father's death wasn't the best, but it could have been a lot worse.

And I guess the timing of his death also saved him from seeing me go on to change careers many more times. He was right when he predicted that I would 'fail' to continue as a practising doctor!

After my father died, my mother and I sat around that night talking about him and having a few drinks and crying. When I woke up the next morning, all the teeth in my mouth were loose. I must have been grinding my teeth so hard during the night from grief that they moved about in my gums. But luckily, my teeth came good on their own.

I think immortality means that you continue to exist in the minds of people who you influenced and who loved you. I'm not much of a believer in spiritualism and other dimensions, but the weirdest thing happened after my father died.

Before he died, I had never written a popular article or journal piece. At school I did my homework easily – except for writing tasks. I just could not write anything creative. For me, writing was a painful process, and the end product was uninspiring. But my father was the opposite. He could write fluently. It was as if he breathed the words out onto the page in an elegant profusion.

After he died, I had this mystical sense of his writing ability being transferred to me. When I inherited his faithful old typewriter, I also seemed to inherit his ability to turn words into stories. I have no logical explanation. Certainly, the idea that there was a writing 'spirit' existing inside his mind that jumped into my mind when he died is totally stupid. But on the other hand, something just like that somehow happened. I still don't understand it, but writing is a connection with him that I really treasure – even if it is completely made up in my imagination. And maybe my move into journalism was based on a wish for him to be proud of me after his death, even if he never approved of my work patterns while he was alive!

My mother and I inscribed his gravestone with the words 'Our Renaissance Man', which in my eyes he absolutely was.

Around 1983, after a lot of discussion, my mother sold her house in Wollongong and bought a house in Sydney's inner west. I moved in with her at Lilyfield, which was not as close to the university as my Glebe squat, but still an easy drive or walk. Soon after, Mary moved in, and then our three beautiful children blossomed forth.

Even at the new house, my mother said she was getting visitations from my father, seeing him sitting on the end of her bed and talking to her. She found this quite comforting.

My mother died more than a decade after my father. She had a slow deterioration with dementia and other brain injuries, and it was painful and sad. In the early stages of dementia, it can be hard to immediately recognise that this is the beginning of a progressive condition. For my mother, dementia plunged her into hallucinations about her dreadful wartime experiences. She saw piles of dead bodies in our courtyard at home and would come in screaming, demanding to know why we had left our children in a bag behind her door. She constantly thought we were trying to kill her. It was extremely distressing and upsetting.

The medical advice we got at the time was that because she had deteriorated so quickly, she was not likely to live for a long time. That wasn't true. For about ten years she lived in a very reduced state in a nursing home. At least in the early part of that, when she was still able to walk, I could take her out to music concerts with her old friends, but over time she became bedbound and nonverbal. After a while, the only way I still interacted actively with her was feeding her food, but especially chocolate, which she seemed to enjoy until the very end.

Mary visited her the night before we were heading off for a short trip to Darwin. She came home very agitated and insisted I go see my mother that night, because she felt Mum was about to die.

By this stage I thought my mum was just about immortal, given how long she had survived in the nursing home without walking or talking and barely eating. But I went to see her anyway. I thought she looked really good, maybe the best I had seen her for ages. It seemed like she took my hand deliberately and, when she stared at my face, recognised me in a way she had not been able to do for years.

I came home and told Mary that she was perhaps exaggerating. Mary looked serious, and said now she knew for sure that my mum was going to die and was somehow rallying to connect with me a final time. I didn't really take Mary's concerns seriously, and we went about packing for our little trip.

The next day we went to the airport and boarded our plane. After the doors closed and everyone was ready to take off, my phone rang. (We hadn't yet been told to turn off electronic devices for the flight.) It was the nursing home ringing to let me know that my mum had finally died.

I was rocked to my core. We had been waiting for this to happen for many years, but somehow I still didn't expect my mum would die. It was confusing emotionally because I didn't want her to die, but I also didn't want her to be trapped in the restricted life she had been living. But now there was nothing I could do to change anything. At least I had gone to see her that one last time, for whatever that was worth.

I officially became an orphan at that moment. For the first half of my life I had no family apart from my parents, so it was strange to have neither of them anymore. Even though by this point I had my own children, it took a long time to not feel very alone after my parents both had died. My parents had always seen themselves as a single unit, separated from the rest of the world. Now I found myself simultaneously 'alone' – but very fortunately, still buoyed up by the family Mary and I had created. On one hand, all ties to my European heritage had been cut. But on the other hand, I had my own partner and children, and was part of an enormous Irish Catholic family that was as varied as you could imagine.

Emotionally, I was so lucky to be swimming, not drowning, in grief, lucky in the support of my family.

What I had left, after my parents died, was my mother's love of classical music and my father's ability to write stories!

NASA applications and media lift-off

Jumping back in time, the year 1981 was a transformational one, not just because I started med school and lost my father but also because our understanding of the universe was changing. Early that year, the US space shuttle *Columbia* was about to enter service as the first ever reusable rocket, with the ability to ferry people up and down from space to launch satellites and do research.

I am in love with space and have been since I was a child. I guess it is another of my true loves! Space is so big and full of possibility. It is exciting, and to me it embodies every sci-fi future that I ever dreamed about.

A Periodic Tale

I am fully prepared to blast off into the unknown depths of our universe if anyone would give me half a chance to go. In the past, as a single bloke, I would have taken a one-way trip into space. Now that I have a family, I still desperately want to go to space, but I would prefer it to be on a return ticket! After all, one in every seventy space shuttle flights has ended in everybody dying ...

Back in 1981, I would have given anything to go up and down on the space shuttle. The upcoming launch prompted me to write a detailed application letter to NASA. It went something like this:

Dear NASA,

I would like to become an astronaut. I have a bachelor's degree in physics and mathematics, a master's degree in biomedical engineering (where I designed and built a machine to pick up electrical signals of the human retina to diagnose certain diseases of the eye), and I am in my first year at medical school aiming to get degrees in medicine and surgery. My skills would surely be useful to you. I am thirty-two years old and very fit.

Yours sincerely,
Karl

Shortly after, NASA sent me a letter typed on a typewriter and signed personally by a human. It basically said, 'Dear Mr Kruszelnicki: First, we already have a full quota of astronauts, and second, our astronauts have to be American citizens. Sorry. Cheers.'

Maybe I was delusional, but I really thought that I had a chance of joining NASA after I finished my medical studies. I certainly thought it was at least worth a go!

If my dad was telling this story, he would describe it as an example of another failure in my life. But despite being turned down flat by NASA, I still saw possibilities. I remained fascinated by the amazing space program and the technology in the American space shuttle and kept reading everything I could about it.

By pure chance I heard that the ABC youth radio station 2JJJ (now just known as Triple J) would do live coverage of the very first launch of the space shuttle. The station was wildly experimental and incredibly hip. Announcers said anything they felt like on air, clothes were semi-optional, and sobriety was definitely not necessary. Once an announcer nodded off at the mic and let the record he was playing continue to go 'tic tic tic' when it got to the end of the grooves – for a good forty-five minutes! When he came to, he back-announced the track in a deep, smooth voice as if nothing untoward had happened. It was the days of ambient music, after all! The audience sucked up the vibe and loved the 'out-there' attitude of the station.

So in keeping with my philosophy that they can't shoot you for asking, I rang up JJJ and offered my services to talk about the space shuttle. I love talking on the phone – always did, always will! – and my call must have pricked the producer's interest because he asked me to come in for a chat. I did, and within a few minutes he invited me to continue our conversation about the space shuttle in one of the recording booths.

Although I didn't realise it at the time, this was a practical test of my suitability to speak on air – a kind of job trial. I love talking and can pretty much do it anywhere, and speaking into a microphone did not throw me off balance at all. What I always try to do is turn what I know from reading scientific literature into stories that people can understand – especially if they don't have all the background training that I do. The initial chat went so well that I was invited to come back and cover the actual launch live-to-air in the studio.

With today's communication networks, you can watch almost anything live, from anywhere, with no prior planning. So it's hard to imagine how big a deal it was back in 1981. They had to organise the audio links from Florida, to LA, to Honolulu, to Overseas Telecommunications in Sydney, and finally back to the JJJ studio in Sydney. But come the night of the launch, the connections were working and everything was on track.

At first, the initial space shuttle launch on 10 April was also going fabulously well – until a problem developed with one of the fuel cells. Fuel cells were essential (or 'mission-critical', in engineering speak) because they supplied electricity for the space shuttle. Without onboard electricity, there was going to be no launch that day.

In the live-to-air studio, everybody looked to me for an explanation about what was happening and why the launch wasn't going ahead. Luckily, I knew about fuel cells, and spontaneously chatted about fuel cells being invented back in the mid-1800s, and that they were a box that would give you electricity so long as you kept on feeding them fuel, usually hydrogen. A fuel cell did the chemical reaction of *hydrogen + oxygen = water + electricity* in a remarkably clean way. The 2JJJ folk were very happy with our unscripted broadcast.

With the launch cancelled for that day, the live broadcast came to an early finish. But two days later, we were there again for the rescheduled blast-off. This time the launch went smoothly, and I got to talk everyone through the whole event.

Afterwards, we were all hanging out in the tearoom when the producer said, 'I really need this cup of herbal tea to clean my kidneys.'

Without drawing breath, I launched into my kidney spiel: 'Actually, it happens the other way around – your kidneys do the cleaning and remove the herbal tea from your blood. And did you know that your kidneys filter about a sixth of a tonne of your blood each day? And at tremendous metabolic cost, they also remove about half a kilogram of salt from your blood. And then, because evolution made a mistake and we are basically fish gone wrong, our kidneys put all that salt back into the blood except for a few grams, which leaves your body in your urine.'

As soon as I got to the end of my little monologue, the producer said, 'We need you for "Great Moments in Science".' I'd accidentally launched myself into a media career – not quite space, but very 'out there'!

I would go into the JJJ studio every few weeks and prerecord a bunch of three-minute science audio stories that we called 'Great Moments in Science', or GMIS. The radio presenters were wild and carefree and unregulated; it was an amazing eye-opener for me. Everyone had stories to tell, and it was such an exciting, buzzy place at all hours of the day and night. I found a new niche that I simply fell into, and I loved it. Amazingly, I even got paid for doing it!

I've been on Triple J practically the whole time it has been running. I'm the most long-term presenter they have! I feel like a reverse of Dr Who: instead of me regenerating, everyone else regenerates around me – they get younger and I get older. It is pretty funny that I have been on the 'youth' radio station now for over forty years!

One thing you have to be able to do on radio is speak clearly into the microphone. But microphone fright is a real phenomenon, and I was hit with it a few months after I started in the job. There was some breaking environmental news, and they asked me to do a live read in the news bulletin. Until now – apart from the initial space shuttle launch – everything I did had been a prerecord. For some reason, I belatedly realised that if you make a mistake live, you can't take it back. I just couldn't do it! In the end I had to prerecord the segment and then it was aired after the news. Luckily this was a temporary problem, and I cured myself by constant exposure to the microphone in live spots, kind of like a desensitisation training for phobias. (I am very phobic about spiders, but I haven't ever cured myself of that fear. I just get someone braver – Mary! – to remove any terrifying beasts. I did once buy a photographic book of spiders and practised touching the spider pictures, but that's as far as I got.)

I learned a lot on Triple J from one of my early producers of the 'Great Moments in Science' stories. He helped me do rewrites and taught me badly needed microphone and editing techniques. He pushed me to always be creative. He told me that every story had to have a beginning, a middle and an end. The simple formula was to start with something amazing to hook the audience in, explain it fully in the middle and then finish with a joke. It's an age-old formula, but it has worked well for me – and for generations of storytellers before me too!

The GMIS stories were attractive enough to get other people's interest. In 1984, a publisher approached me and asked to turn the stories into a fun book. What did I think? I thought, sure! I had to do some editing of the radio stories to make them suitable to a book format, then the publisher organised captions and illustrations, and a few months later my very first book appeared, a slim blue paperback called – you guessed it – *Great Moments in Science*.

I wanted to help publicise my book, but I didn't know how to get the word out about my new little literary baby. I was still a fourth-year medical student in the hospital at the time, and in the wards I would see patients stuck in bed all day watching daytime TV. (And as it turned out, so did the nurses and police and a whole lot of other shift workers.) And the most popular daytime program then was *The Mike Walsh Show* at midday on Channel Nine. Seeing everyone staring at the same TV show every day made me wonder if that might be a way to get some good publicity for my new book.

So again, I just got on the phone. I simply rang up *The Mike Walsh Show* and said to the producer who picked up the phone, 'I've got a new book on popular science, and one of the stories is about why women synchronise menstrual cycles when they live together and how you can use this information to tame a wild dog.' He loved my pitch and invited me on. A few days later, after being picked up in a fancy hire car, there I was on live TV telling the same story to the audience. It was the first of some regular spots on the show to come, though not all of them went as smoothly! Over time, Mike Walsh moved on, but the show continued as *The Midday Show with Ray Martin*.

> **The Dunning–Kruger effect**
>
> Maybe you've heard of a psychological phenomenon known as the Dunning–Kruger effect. Simply, it is when you know only a tiny amount, but for no good reason you match that lack of knowledge with an abnormally high amount of confidence.
>
> When you gradually get more knowledge and realise that there's more to this topic than you first thought, your confidence about how much you understand decreases proportionally. As you get even more knowledge, your confidence decreases to rock bottom, and you get to thinking, 'I'm never going to understand any of this!'
>
> But as you learn even more and head closer to properly understanding the topic, you begin to cautiously increase your confidence. You move from 'I'm never going to understand this' to 'Oh, it's starting to make sense'.

> Ideally you finally finish up with more knowledge than you had at the start and a more reasonable level of confidence. You might say something like, 'Look, I know a little bit about it, but it's a lot more complicated than I can explain.'
>
> During my early years in science talkback radio, I suffered from the Dunning-Kruger effect to varying degrees at times. But if it has taught me anything it's how to be very happy and confident about saying, 'I don't know!'

One particularly memorable spot on *The Midday Show* almost got me fired! But it ultimately led to my career blossoming, because it made me realise you had to speak in your own voice and let your personality shine through.

I was working full-time as a junior doctor at Concord hospital, but my immediate boss was happy for me to duck off to do television spots – as long as I still fulfilled all my hospital responsibilities, didn't stuff anything up and didn't increase anybody's workload.

Fridays were best for me to do the TV gigs. I would leave hospital just before lunchtime, drive quickly to Channel Nine, go on air and do my live spot, and eat whatever goodies they had in the green room for my lunch. Then I would head back to the hospital and finish my afternoon shift. The patients very quickly cottoned on that the guy treating them in the morning and afternoon was also the guy they were watching on the Friday *Midday Show*, which at the time was the highest-rating show on daytime TV in Australia. My surgeon boss was amused that patients were more impressed by me (who was very junior but appeared on TV) than they were by him, who had literally saved their lives in the operating theatre.

On this one particular occasion, the host, Ray Martin, said, 'We'd like to do something about feminism as a special show.'

Now sure, I'd heard about feminism, and I was a hundred per cent on board with equal opportunity expectations, but I really hadn't read any feminist ideology at all. So I said to Ray, 'Look, I roughly know what it is. But I really don't know anything about the theory of it.'

And he said, 'By an amazing coincidence, I can help with that. I've got a book. Here, take this autographed copy and read it and come back for the show on Monday. Keep it, it's yours.'

Now the book was by somebody you've quite probably heard of named Germaine Greer, and the book was her long and densely intellectual text *The Female Eunuch*. Not a light weekend read in any sense!

As it turned out, I spent the weekend having too much fun and I didn't even open the book. So when I turned up for an unusual-for-me time slot on Monday, suddenly everything went wrong – on every possible level.

First, I hadn't read the book.

Second, I wasn't on just for my regular five-minute segment but for the entire show. And it turned out that not even opening the book was a major issue, because the whole show was about feminist theory.

Third, the show didn't run just for the regular one and a half hours but for two hours. Oh my God! How do I survive for two hours, with nothing to say?!

Fourth, the reason it was running over into two full hours was because all the famous feminists in the known universe were on the show, including Germaine Greer herself, Naomi Wolf and a host of others! Ray and I were the token males.

As you might expect, it started off badly for me. Normally the first segment was the drawcard and ran for longer than the later segments, in order to drag the audience in and make them want to watch the whole show. But for that entire fifteen-minute segment, I said absolutely nothing! I was thinking, 'Well, I don't know anything, but at least I can keep my mouth shut.'

In the break between the first and second segments, Ray Martin leaned over to me and said, 'You okay, Karl?'

Lying through my teeth, I answered, 'Yeah, I'm fine.' But deep down I was panicking and thinking, 'I've got to get through another hour and forty-five minutes of saying nothing on live TV.'

But I calmed down and tried to listen to the clever people talk about feminism, Gradually, half an idea started to surface in my mind, not from doing any reading but entirely from hearing and concentrating on what they were saying.

Now at this stage I want to point out that when you're doing live TV and radio, at some stage you have to bypass the 'internal censor' in your brain that stops you from saying stupid stuff. You don't have time to think about what you say before you say it. Dead air, when no one says anything, isn't very entertaining!

I nodded to Ray because I could feel this idea gradually building up inside my head and because he was looking at me and anxiously signalling, 'Come on, do something.' So he flicked a Dorothy Dixer question to me, along the lines of 'What do you think, Karl?'

A camera slowly zoomed in on me, which meant my face filled half a million TV screens around Australia. To my absolute horror, I heard myself saying very clearly, 'Well you know, Ray, what I've picked up here from listening to Naomi and Germaine is that us blokes have to stop thinking about penile penetration, and start thinking about vaginal engulfment.'

OMG, what had I just said?!

I had never put those two words together before, nor had I even heard them linked before! I'm pretty confident I was saying those words for the first, and possibly only, time on live TV. Ray looked at me, aghast, and even he could think of nothing to save me. In fact, nobody said anything at all. There was dead silence. You could hear a pin drop in the studio.

Suddenly, the studio monitor showing the live TV broadcast signal going to air went to black. Somebody had decided to pull the plug. Going to black is a very bad thing!

Now I had been in TV long enough to realise that the important stuff is the ads. They generate the income. The program is just there to keep you watching until the ads come on. You can't go to black when the advertisers are paying $5000 a minute.

In a panic, no doubt super desperate to put anything other than me on the screen, somebody in the control room pulled up the first ad they could find to fill the airtime. It was for sheep dip! By the way, that ad had previously been shown only to country audiences, and never to city folk.

Then some rent-a-cops came and dragged me off the stage and I went very willingly. I wanted badly to escape!

I thought things couldn't possibly get any worse. But then they did.

The switchboard lit up as people started to ring in, saying, 'Hey, I just understood what that guy said. I'm beginning to get this feminism stuff.' And so, they dragged reluctant me back on the show. I had to stay there and try to wing it for another hour and a half!

(At least, that's my memory of it ...)

My first story on climate change

I wrote my very first story on climate change in 1981, way before climate scientists started getting death threats for just doing their job and reporting the facts. Back then, climate change was usually referred to as the greenhouse effect or global warming.

I've kept writing about it ever since because I believe I have an obligation to speak publicly on this issue, to counter the lack of interest and outright lies that have turned this debate from a factual exchange into a campaign of misinformation and cover-ups.

Having grown up under the cloud of nuclear holocaust, I understand the similar despair that climate change is bringing to the community! Climate change is to Zoomers and millennials what the nuclear threat and the Vietnam War were to my generation.

In 1973, the world's largest reinsurance company, Munich Re, had already increased insurance premiums because of extreme weather events such as floods resulting from climate change. They were ahead of the scientists in accepting the climate data because it was costing them money. Nothing personal, just business. And all the other insurance companies around the world followed suit.

By 1977, the chief scientist of the oil company Exxon, James Black, was on board. He told management, 'There is general scientific agreement that the most likely manner in which mankind is influencing the global climate is through carbon dioxide release from the burning of fossil fuels.' As a result, Exxon teamed up with other fossil fuel companies and between 1977 and 1990 undertook world-best research into climate change. How's that for surprising!

By 1990, the climatologists were convinced that the data they were recording showed a trend of overall heating of the planet that was way beyond natural climate variations.

Unfortunately, in 1990, the fossil fuel companies completely changed direction and instead set up lobby groups to thoroughly embed fossil fuels into all aspects of society. To do this they funded massive disinformation campaigns.

Carbon dioxide levels in the atmosphere have risen by about 140 parts per million (ppm) since 1750. We've seen records for maximum temperature levels broken over and over along with multiple extreme weather events. Even so, fossil fuel companies have been denying their role in climate change like crazy!

The other craziness has been the way that regular climate change scientists have been vilified in deliberate smear campaigns, as well as having to live with eggings, legal arguments and even death threats.

I throw everything I've got at this topic to keep it in the public eye. I hope eventually that the political decision-makers will act to prevent runaway changes to our planet, which would make life really complicated for the next generation!

I didn't know it, but around the same time (year four of medical school) that I was learning how to be a daytime TV guest on *The Midday Show,* the ABC was starting up a brand-new popular science television show. My publisher sent a copy of my first book over and they liked the stories, so ABC TV got hold of me.

I went in for a screen test and was genuinely surprised that there's nothing more than a bit of makeup needed to create a more photogenic version of me!

Next thing I knew I got a phone call inviting me to be one of the three presenters in the first series of a half-hour show called *Quantum.*

I was in a bit of a conundrum. On one hand, it was an attractive offer. But on the other hand, I really wanted to keep going with my study and head into fifth-year med so I could become a doctor in the hospital system treating patients. I was thirty-six years old and felt like I needed to put my nose to the grindstone and get this medical degree under my belt!

The next day, sitting around on the lawn at hospital during a break, I mentioned the TV show offer to my fellow students and that I had decided to reject it. They all instantly said I was crazy, and I should take a year off to just go for it. I suspect that if you are about to become a final-year medical student and looking down the barrel of the last big exam block, pretty much any other option is going to look glamorous and attractive!

So I bowed to public consensus and accepted the offer from *Quantum.* I went from a full-time medical student who moonlighted as a TV guest to a full-time television presenter on hiatus from medical studies. It was the beginning of my TV work taking off along with my rise to micro-stardom in the World of Media!

The story I'm most proud of in my *Quantum* year was helping defuse the hysteria about a new disease called AIDS.

At the time, a three-year-old girl and her family were being ostracised by their friends and neighbours because the girl had acquired AIDS from a blood transfusion in 1982, before the virus had even been identified. She was treated terribly, and eventually the family moved to New Zealand.

Back then, there was no cure for AIDS, and people died from it. That was frightening! But it was virtually impossible to get AIDS from casual contact or even from kissing.

How did I know it was impossible?

Well, for a *Quantum* episode I spoke to the best medical experts and worked out that to get AIDS from kissing somebody with the virus, you would have to swallow eight litres of their infected saliva! I held up an 8-litre container of a lovely pink liquid to demonstrate how little risk was posed to the general community from people with AIDS.

That was probably my first public health story – to be later followed by others on microsleeps, vaping, alcohol, vaccinations and many more.

A few years later, the word 'quantum' got me fired from a job before I even started!

I had been hired by the Channel Seven science and technology program *Beyond 2000* as a reporter. At the welcoming party, the executive producer said something like, 'I'm so glad we have Karl on board – it's a really big quantum jump.' Then he invited me to the stage.

I stood up and said, 'Thanks, everyone, but let me make a slight correction. Quantum jumps are tiny – not big at all. In fact, they are among the smallest things we can deal with.' Then I sat down.

I didn't realise it, but the executive producer didn't like anyone to disagree with him. Others told me his face went black with anger and he left the room as I finished speaking.

The next day I didn't have a job at *Beyond 2000* anymore, so I guess he was correct – for me, that job really was a quantum jump!

But *Quantum* was an absolute blast, and I had so much fun. I picked up more new knowledge, but as the first season came to an end, I knew in my heart that I was going back to finish medicine.

1988–92

Why I became a doctor

Intellectually, I became a doctor because I just loved to pieces the bizarre complexity of biological systems. I mean, we still don't understand how a bunch of chemicals get organised into us humans who can reproduce ourselves!

I jumped in what seemed to me natural and organic stages: from being a physicist, to being an academic and biological researcher in Papua New Guinea, to being a groovy hippie/filmmaker/taxi driver/roadie living life to the max, to working in a hospital as a scientific officer, to building a diagnostic eye-disease machine for Fred Hollows, to studying medicine. I felt some kind of altruistic vibe in my underlying desire to study medicine, thinking that in some kind of hippy-dippy way, I could make the world a better place. Usually, I'm more interested in trying to understand the world *around* me than *inside* me, so a deeper understanding about why I decided to become a doctor didn't come until some years later on the job.

A Periodic Tale

I finished studying fifth-year medicine in 1986. In 1987, I spent my first year as a Real Medical Doctor. I was a lowly intern, at the very bottom of the doctor chain. During that year I worked in half a dozen fields (including emergency, gastroenterology, respiratory surgery and cardiology) at different hospitals around Sydney, spending a few months in each speciality. In 1988, I went up one micro-step from intern to resident. Again, I rotated through many different medical subspecialities in different hospitals, learning so much and so quickly.

And then, in 1989, the Big Change: I became a resident at the Royal Alexandra Hospital for Children in Camperdown, commonly known as the kids' hospital. I liked regular adult medicine, but this was heaven on a stick. And on one magic night at the kids' hospital, I accidentally found out why I became a doctor.

Around 10 pm on a winter night, a four-year-old boy was brought to the emergency department by his parents. The first thing I saw was parents that were both very upset and tired. The kid – classic spot diagnosis – had some kind of respiratory disease, possibly even pneumonia. (A spot diagnosis is one you make within five seconds of looking at the patient. For example, a person's walk can instantly indicate a stroke, or Parkinson's disease, or polio.) He looked tired too, with bags under his eyes, and his face was flushed with fever. He was breathing much too quickly, and with each breath he sucked in the area above his collarbones as he struggled to get oxygen.

As soon as we sat down, the parents raced to tell me the story in great detail. They had taken very comprehensive notes – this was their only child, and they were worried sick. Their kid had had rolling fevers (carefully documented), which had been going on for weeks. He wasn't eating. But every single time they took him to a doctor, the fevers would go away and the kid would look fine, so there'd be nothing for the doctor to find!

But now they had arrived at the hospital with their child showing the symptoms right here, right now, in front of a doctor – me. I picked up their underlying current of fear that this was more than a simple infectious disease and that perhaps something was very seriously wrong.

I gently suggested that I should now do a very thorough and comprehensive physical examination, but first I went into the corridor and arranged for three cups of tea and an ice block – I could see this was going to be a rather special and long investigation, and we always kept ice blocks for the kids with fevers who needed fluids. It was a bit unusual in the emergency department to take time out for a cuppa, but I felt this family needed the extra support.

To treat the child, you also have to treat the family. The thorough physical examination was to reassure the parents that their child was being completely reviewed. And the tea and ice block were to help bring the stress levels in the room down.

Yep, the kid had a fever, and yep, he had stopped putting on weight over the last few weeks. I did the examination and practically everything was normal – apart from his respiratory system. The parents' faces relaxed as I checked the child over, like they were finally being taken seriously and weren't imagining things.

I said, 'I'm very confident that this will all have a happy ending. But to be a hundred per cent sure, I need to order a chest X-ray.'

The parents happily went off to radiology with their child.

I was with the next patient when I heard a doctor say, 'Hey, Karl, you hit the jackpot!' I looked at the child's X-ray that he was peering at in the light box, and there it was: a little well-defined pneumonia down the bottom of one lung.

I brought the family back into the little cubicle and said, 'I've got a tiny bit of bad news that is completely wiped out by a huge amount of good news. The bad news is that you were quite correct – there *is* something wrong.' I showed them the little pneumonia on the X-ray. 'We need to admit your child overnight to treat him with antibiotics, and in a few days it will be as though the pneumonia never happened. And then you can get back to enjoying your wonderful child.'

Being able to give someone news about a treatable and fixable medical condition is quite joyous. I unexpectedly found myself choking up a little and having difficulty speaking. The parents burst into tears, and I found myself crying too.

And at that exact moment, I suddenly realised, deep down, for the very first time, why I became a medical doctor. It was to help liberate people, where possible, from the burden of disease.

Insights into people's lives

Being a doctor can also give you strange insights into people's lives.

One night, around 8 pm, two parents and their three-year-old came into the emergency department at the kids' hospital. At first glance, no emergency was apparent. The nursing admission notes said '?deafness', which was a little unusual in a three-year-old.

The story was that their housekeeper reckoned their child was at least partially deaf, but they thought that was impossible – wasn't it?

I asked them to explain about the 'deafness' to me, but they said they couldn't. Their child spoke only Spanish, not English, while they spoke only English, not Spanish. What?

It turned out they were a well-known and wealthy couple who spent virtually no time with their child. The child had been brought up from birth by their live-in housekeeper, who was Spanish-speaking. The child spoke virtually no English but, of course, was fluent in Spanish. The housekeeper had been worried about the child's hearing for some time, but the parents had ignored her concerns. Tonight, in the late hours, in a sudden panic, they were finally seeking medical care.

This problem did not fit into the emergency pathway, but I did refer them to the audiology unit at the hospital for a hearing check-up as a next step.

It left me feeling very odd that the parents spent so little time with their three-year-old child that they didn't even speak the same language, and that they hadn't noticed themselves that their child could be deaf!

Gastroenterology versus neurology

When I embarked upon the Giant Ocean of Knowledge that is medicine, I had no idea which subspecialties I would love.

I was surprised to find myself really enjoying surgery. The surgeons' mottos included 'A chance to cut is a chance to heal' and 'Nothing heals like cold, hard steel'. What I liked about surgery was the directness of the surgeons themselves – what you saw was what you got! I also especially liked that sometimes a person's medical problems could be very quickly resolved with surgery.

I also found to my surprise that I really enjoyed gastroenterology (or 'gastro'). Gastro covers everything from the mouth to the anus. I was surprised and delighted to find that if you had good knowledge, and could also communicate with the patient, you could work out the diagnosis without even laying hands on them about 60 per cent of the time. In another 20 per cent of cases, to make the diagnosis you also had to lay your hands on the patient (for example, to feel for a lump in the abdomen or groin). Only in the remaining 20 per cent of cases did you have to do 'investigations' (like blood tests, imaging or scoping). This made gastro intellectually satisfying for me, and as a bonus, the abdominal conditions were often treatable.

Interestingly, roughly the same breakdown of statistics kind of applied to neurology. But unfortunately, at least in my brief experience, while you could make an accurate diagnosis in neurology, the patient would often be stuck with something that could not be fixed (like a cancer in the brain).

Fred's foot

Some cancers can spread silently, such as kidney cancer. This is what happened to my mate Fred Hollows, who was so sadly struck down before his time by this.

I loved knowing Fred. He was an inspirational medical doctor and an exceptional humanitarian. He also helped me enormously in my career. And he operated on my mother's cataracts at the basic Medicare rate when he could have charged a lot more. Fred did so many good things for the community in general, both within Australia and overseas. He was what you would call a rough diamond, which means he was a good-hearted person who tended to drop the f-bomb a lot!

I kept in contact with Fred after my biomedical engineering days and popped in to his place from time to time. I took my beloved fully kitted ex-military C304 four-wheel drive around, after it was all done up for travel anywhere and everywhere. I knew that Fred would be interested, because he had done a lot of outback travel. I wanted to show off my handiwork to him. He was suitably impressed, and especially liked the outdoor shower head that got warm water from the on-board tank, via the heat exchanger.

I dropped in on him one day for a cuppa. 'How are you going, Fred?'

He was characteristically blunt. 'I'm fucked. I couldn't find my bloody left foot this morning.'

Fred had been diagnosed with kidney cancer, which is one of those cancers that can spread through the body. So I kind of knew what he was getting at but asked, 'What do you mean?'

He said, 'I've got bloody Gerstmann syndrome.'

My heart sank. Gerstmann syndrome is a rare disorder that is due to damage, sometimes from a cancer, in the part of the brain near the parietal and temporal lobes. The four classic signs of Gerstmann syndrome are inability to write, inability to work with numbers, inability to distinguish between the left and the right parts of the body, and, in its classic form, finger agnosia. Agnosia means to not know, and typically, people with Gerstmann's can't recognise their fingers. But foot agnosia?

Fred went on. 'Yeah, I tried to put my sock on this morning. I could put a sock on my right foot. But then I couldn't put the other sock on my left foot. There was something at the end of my left leg, but it was all wrong. I knew theoretically it had to be my foot, but I couldn't recognise it to put my sock on it. It took me a while, but I worked out that I've got Gerstmann's. So that means I've got cancer near the parietal lobe and I'm really fucked.'

Unfortunately, Fred died shortly after.

Fred was just 63 when he died, and he died long before he deserved. He still had young kids, plenty of work to do, and lots of life to enjoy. But he didn't get an option. None of us do!

Medicine outside hospitals

I worked as a medical doctor in the hospital system for less than ten years, much of it part-time at the kids' hospital, so I don't know a lot about adult medicine. But at least once a year, I get called on to help in a medical emergency. This is especially common when I'm flying on an aeroplane. Some medical people on planes refuse to admit they are medical doctors. I guess that these doctors might feel that they are in holiday mode, so they need a proper break. But my attitude is that I have a duty to pay back the cost of my medical education to the community in some way. I am also reassured by the Good Samaritan Laws, where if you offer to help within the limit of your skills, then you are protected from personal liability.

On average, there's a medical emergency on one in every 600 flights, or another way of looking at it is there's roughly one medical emergency for every 11,000 people travelling on aeroplanes. But for some reason I seem to buck the odds and get more than my fair share.

My record was in 1999. I took 141 flights in that calendar year and had five medical emergencies to deal with – four lived but one died. Since then, I've helped in roughly one airline medical emergency per year, and nobody has died since.

In order of decreasing frequency, the symptoms that doctors on planes are called about range from feeling faint, difficulty breathing, nausea or vomiting, chest pain, seizure, abdominal pain or some sort of infectious disease. And yes, that fits with my personal airline medical drama experiences.

I've also realised over the years that when I talk on the radio about certain conditions, people listening might remember the story if and when those symptoms happen to them, so inadvertently I 'treat' them too.

A good example is when I've spoken about detachment of the retina. The retina is a layer roughly 0.3 millimetres thick on the inside back of the eyeball. (The eyeball itself is about 24 millimetres in diameter.) Sometimes part of it can peel away and wither, usually following some sort of fall or minor head injury. If the retina peels away and detaches, you might experience a symptom like a curtain falling across part of your vision in that eye. This is a medical emergency, and you need treatment as soon as possible.

One risk factor is a slightly bigger eyeball, as usually happens when you have to wear glasses for long vision. The physics of surface tension tells us that with a bigger eyeball, the retina is not as tightly adhered to the rest of the eyeball, compared to a regular-sized eyeball. Really short-sighted people are at higher risk of retinal detachment – one in twenty! The overall background risk for the general population is one in three hundred over a lifetime.

The treatment for retinal detachment is relatively straightforward: get yourself to an eye hospital and get your retina repaired within six hours and all will be well. But if you delay, you can lose vision on that part of the eyeball for the rest of your life.

Once I got feedback from a teenager who had gone on a sports camp, several hours out of Melbourne. She fell heavily and received a blow to her head, and then noticed a typical curtain-like loss of vision. She went to the organisers and said, 'I've got a detached retina. You have to take me to the eye hospital in Melbourne right now.'

Melbourne was several hours away, so the camp organisers were understandably not immediately on board and asked, 'What are you talking about?'

She told them, 'I heard Dr Karl on Triple J say that if you have symptoms like mine, it means you have a detached retina, and you have to go straight to an eye hospital to get it fixed.'

And so, trusting me via her, they took her to hospital and her vision was saved.

Another person in California also heard me talk about detached retinas on Triple J. By chance, he called his uncle, also in the United States, after the show, who told him about a weird experience just a few hours earlier when a curtain-like block fell across one side of his vision. He told his uncle to go to the nearest eye hospital, and that was also a happy ending.

In another instance, someone who suffered a ball injury remembered a story I told on the radio that saved her life.

The bones in the skull come in different thicknesses, and the thinnest of all are the temple bones on each side of the head. Inside the skull, the middle meningeal artery runs in a groove on the temple bone. If there is an injury on the side of the skull, such as from a flying ball, the temporal bone can break and buckle inwards. At that instant, the artery is torn apart. The bone has some elasticity and can spring back to its original position, but the artery is still torn! So while on the outside there may be no bruises, on the inside the artery is leaking away.

In the classic case, a person will be hit in the temple and then briefly fall unconscious. They wake up shortly afterwards seeming very clear-headed – the so-called 'lucid interval'. Everyone is asking how they are. They feel fine. But invisibly inside the skull, the middle meningeal artery is slowly leaking blood into the space between the brain and the skull.

At the bottom of the skull is a big hole with a Latin name, foramen magnum, which literally means 'great hole'. As the leaking blood builds up in volume inside the skull, it pushes the whole brain downwards. The first part of the brain to push into the bony rim of the foramen magnum is the area of the brainstem that controls your heart rate and breathing – which begin to slow down. So after a little while, the person who has been hit in the head says something like, 'I'm feeling a bit sleepy. I will just have a little lie-down.' And the bystanders nod sympathetically, saying, 'Yeah, that makes sense. It was a big knock to your head.' Left untreated, the leaking blood continues to push their brain downwards. They fall comfortably 'asleep', and then their breathing and heart stops, and they die.

What's needed is an urgent operation to drain the blood (which will relieve the pressure on the brain stem) and to also repair the leaking middle meningeal artery.

This operation to relieve pressure inside the skull – trepanation – has been done for about 10,000 years. This was known to our ancestors in the Pacific and South America.

Historically, with trepanation, the flesh and bone over the artery were cut with something sharp, such as an obsidian knife in South America or a sharp piece of coral in the Pacific. The two cut ends of the artery were tied shut with spiderweb, which has its own inherent antibiotic properties. How did they join the outside skin together? Many different ways. One especially elegant one was to pull the flesh together with fingers, then push a bull ant with large pincers up against the incision. The pincers would close tight and hold the flesh together, then the ant body was twisted off leaving just the pincers behind.

Until the late 1800s the Pacific and South America had a much higher success rate and lower post-op infection with trepanation compared to Europe. That's when the Europeans began to understand the importance of clean operating fields and sepsis.

In the late 1990s there was a case of trepanation in Central Australia, where a passing telecommunications technician used a drill and a hole saw to relieve the pressure from inside someone's skull.

Getting back to my particular case, a teenage footballer had received a blow to the side of her head. She went unconscious and then became aware again. She had heard me talking about this 'lucid interval' on radio, and began to wonder if she had a bleed in her middle meningeal artery. She felt herself getting tired and said to her parents quickly, 'Take me to the closest hospital and tell them Dr Karl says I've got middle meningeal artery bleeding.' They took her straight to hospital where she collapsed at the doors of the emergency department!

A Periodic Tale

Her parents blurted out the magic words, and the operation was done straightaway to relieve the pressure on her brain. Happily, she lived without any further complications!

It makes me so pleased to hear these stories with their upbeat endings. I feel confident that I made a good choice going into the media. Working in the media, I can reach so many more people with health and science stories, than I ever could have done by working as an individual doctor seeing patients one at a time.

Journalism has the potential to reach people and do good overall by airing solid, reliable, informative stories. But the media can also do bad, by running controversial unbalanced stories, in the quest for audience attention. In the 1990s, the TV program *A Current Affair* was running a sensational campaign against vaccinations in general, and especially the whooping cough vaccine, presumably because it made for good ratings! I was working at the kids' hospital at that time, and in the aftermath of their anti-vax campaign, vaccination rates for whooping cough did fall. What followed in the hospital was little babies coming in with the terrible symptoms of whooping cough and being unable to breathe. Eventually one of these little sick babies died from a disease that can, in the vast majority of cases, be prevented by vaccination.

I really laid the blame for the drop in vaccination rates, and the subsequent increased infection rates with whooping cough, in the hands of those anti-vax media stories.

I wanted to be able to directly counteract that kind of dangerous misinformation with my own media platforms, to let me reach many people directly. So, one more time, I shifted gears.

I left what was probably the most satisfying job of my entire life – being a doctor at the kids' hospital – in the hopes of making a difference to the lives of so many more people than I could treat individually.

1990s

Pregnancy and parenting

Mary and I chose to have children before we decided to get married. For us, having children was the greatest commitment we could make to our long-term relationship – way bigger than buying a barbecue together, that's for sure! And it worked out for us in the happiest possible way.

Love is not a zero-sum game, in which only one person can win and the other must lose. Love can keep on growing out of nowhere, and in doing so, it breaks the laws of the universe (love is energy, and you're not supposed to be able to get energy from nothing!). Having children created lots more love in me – for my children, and of course for my favourite organ, the uterus!

The uterus is my favourite organ because every single human developed in one. It is usually the size of an adult's clenched fist, but when carrying a baby it can expand to the size of a shopping bag – and then shrink down after delivery back to its original size, but without any wrinkles! The Triple J team once made me a celebratory uterus cake covered with shocking red icing representing its bloody lining. The cake even included ovaries complete with ovarian cancers, as both the literal and proverbial icing on the cake!

I went into parenting totally blind. Being a father wasn't something I had ever really thought about. I wish in the early days that I had a better grasp of it, and didn't see the daddy role as something to squeeze in after I finished work.

Very few of my close male friends appeared to have any grasp of how to be a functional or long-term parent, so there were no inspiring father role models to be found there. Some friends had no kids, some had multiple kids with different mothers, and some had no interest in knowing their kids at all. One had a child with a Russian millionaire mafiosi princess – through a series of extraordinary events, in which he was a little pawn – and was fighting physical and legal battles just to see his kid.

Even though I had already graduated as a doctor, I found myself genuinely asking Mary how we had managed to get pregnant and have a baby together. It just seemed so incredible! And that was especially true for me because I had always thought my fertility was low. Way back in first-year med, I had offered to donate sperm, but the fertility clinic turned me down because my sperm count was microscopic – it was almost low enough for me to feel like I could give each and every one of my sperm an individual name!

A Periodic Tale

I was forty in 1988 when our first child was born. For some months after he was born, we couldn't agree on his name. I wanted to call our son baby Karl. Mary thought that was just hysterical (and for many reasons she was dead right!). Eventually she came around and agreed to name our first bonny child Karl Alexander. Baby Karl grew into Little Karl, and he's now a full-blown adult a head taller than me, but we still call him Little Karl. There have been so many ridiculous phone conversations when people ask for Karl and we say, 'Big or Little Karl?' But they can't imagine how Little Karl could possibly be the name of our adult child who stands two metres tall! (I've refused all helpful suggestions that we switch to Old and Young Karl.)

I now feel slightly sorry for having burdened him with my name. It works both ways, though. Sometimes his mail comes addressed to Dr Karl (which is me, not him), but other times when he has made some personal donation to a community radio station fundraiser, the announcers call out a thanks to me for the recent donation, when that donation actually came from him!

At first, I was incredulous about being father. I changed our answering machine message to the famous chorus from Handel's *Messiah* ('For unto us a child is born, unto us a son is given') and ran around singing that beautiful chorus all day. He was a tiny, perfect creature who magically turned up in our lives. I used to read in bed with our little baby lying asleep on my legs and feel such incredible emotions.

I remember one of my friends telling me that he left the mother of his child because she 'changed' after having a baby.

I didn't understand how *anyone* could stay the same after having a baby; for me, it was a life-changing experience! I could not bear to be disconnected from my own child's life.

I stayed home from work at the hospital after the birth of our first child so Mary could go back to finish the further medical training she was doing, which meant she was working and studying at the same time. Baby Karl was six weeks old at the time, so she wanted to keep on breastfeeding. Luckily, she worked in a hospital and had access to practical things like breast pumps and fridges, but she didn't get a lot of support with wanting to express milk at work. I don't think it has become much easier these days for breastfeeding mothers returning to work and trying to express breastmilk.

I loved feeding our baby the bottles of expressed milk. I did all sorts of calculations about the volume of milk that he drank and the percentage of Mary's weight that she was expressing every day. I worked out that in one calendar year, Mary fed Baby Karl roughly two and a half times her own body weight in breastmilk.

I also got worried about the fact that newborn babies don't shiver to get warm. Instead they burn brown fat – but I didn't know this at the time. So I kept our house superheated for the first month, until I came to my senses. It was winter, but you could basically be walking around sweating profusely in just your undies at any time in our home, because of my obsession with keeping the house temperature at something very tropical!

I also resorted to a nanny service on those days when there were just too many things to get done while caring for a baby!

I think you have to be flexible about babies, and the more adults you have to help with each child, the better.

Two years later, our next baby, Alice Salomea, was born. We went with the family to travel in remote areas along the Canning Stock Route in Western Australia when Alice was only a few months old. She had huge eyes, and in every baby photo of her there are flies stuck in the corners of those gorgeous baby blues! The flies were one of the few downsides of camping with small children.

Alice was fully breastfed, which made it easy to travel and not worry about bottles and sterilising. I confess that sometimes I got a bit morbid and found myself wondering how I would look after baby Alice if 'something happened' to Mary. We were so far from civilisation and getting our drinking water from wells – even when there were dead birds and lizards floating in them. On the Canning Stock Route, we saw only one other group of travellers in a month of driving. So my extreme back-up plan was that if Mary died, in my heartbreak I could save baby Alice by feeding her the mushed-up brains of some small creatures mixed with water. (I had read about this as an option in an anthropological article relating to Indigenous life in the Australian deserts. Who knows if it was accurate, but I liked being mentally prepared.) This mixture would hopefully have properties similar to breastmilk and buy me time to get Alice out of the desert without her dying. Honestly – I never thought about just taking some formula with us in case of an emergency. How convoluted must my brain be! Luckily for all of us, that last resort never became a necessity.

Once again, I was amazed by breastmilk and how it adapts to the needs of the baby. Our little daughter was able to survive entirely on breastmilk in daytime temperatures up to 48 degrees Celsius. I remember the first time we took her into an air-conditioned shop, she actually turned blue from head to toe! She had been in outside ambient temperatures of more than 38 degrees practically her whole life, so the air conditioning must have felt like she was being put in a deep freeze.

Our third baby, Lola Scarlett, came along eight years later. It took a while for us to get pregnant that time, but the break between the kids worked well because none of our babies slept at night. I don't think our first two babies slept through the night until they went to school!

Mary and I used to split the night to get some sleep. We arbitrarily called my turn night-time, which ran until 2 am. After that, it was morning, and Mary took over. That was well before internet streaming, so I watched a lot of terrible television in the post-midnight hours while trying to get a baby back to sleep – typically black-and-white war movies with ads for infinite steak knives and 'abdominisers' in the program breaks! Otherwise I put the babies in the car and drove around and around. (Thank you, years of taxi training!) I drove to Wollongong some nights and Manly on others.

One night I drove baby Lola all around the construction site for the upcoming Sydney Olympic Games. I was cruising around slowly as Lola slept, checking out all the new things, when out of nowhere, huge black security cars descended on me from every direction, hemming me in and forcing me to brake. As I wound down my window to speak to them, Lola woke up and started crying! All I could say to the security guards at that point was, 'You woke the baby!' Sheepishly, they let me drive away with a crying child, apparently no longer concerned that I was a terrorist mapping the new venue for an attack.

During birth, Lola's shoulders got a bit stuck, which made the delivery difficult and dangerous. She was vividly purple when she was born from the birth trauma, and her baby photos are quite shocking! Luckily, she recovered quickly, and her skin colour has never again been such a lurid shade.

Being a parent has truly helped me to stay feeling young. It is a cliché, but my kids really have taught me so many things, and I love the clarity children bring to the table. For someone like me who would honestly be happy living entirely on random praise and never dwell on what I could do better, the voice of truth from a child is a reality check! It was my daughter Alice who complained one morning to me that my breath smelled 'like a bum'. I never again came to breakfast without brushing my teeth and cleaning my tongue.

Another plus is that babies hone your reflexes better than any sport I ever tried. You get so good at catching things they fling from the highchair or bump off the table unexpectedly. I think (but I'm not sure) that babies temporarily reverse the inevitable deterioration in reflexes that happens with increasing age. After all, the sample size is one (me) and I don't have a control group – and the plural of 'anecdote' is 'anecdotes', not 'data'. Maybe there is a peer-reviewed paper on this ...

I was awarded Australian Father of the Year in 2003, but I'm not really sure why. As the recipient, not a lot was required apart from turning up to accept the award and then turning up the next year to pass it on to the next recipient. It did actually make me reflect on how I wish I had been a better daddy, and I'm still working to be that! I really do wish I hadn't made the kids cry when I was 'helping' with their maths homework, and I wish I hadn't found it so hard to let them do things at their own pace – like picking up balls at tennis, or emptying the dishwasher – and I really wish I hadn't threatened to throw Lola's school shoes away in a fit of temporary madness as misguided punishment for being 'annoying'. I was breaking my own hard-won rule (from *The Godfather*) of never making a threat unless I was prepared to carry it out immediately! Of course, she just laughed at me, because she knew I was bluffing.

Lola and my nieces are my TikTok gang – I need them to keep me up with the kids on social media! Lola is also in charge of my beauty regimen, which used to just consist of sunblock with sorbolene cream as a moisturiser. She now attends to my eyebrows and keeps them at a minimal level of bushiness!

I've even got a granddaughter, Gemma; young nieces, Frida and Lucia; and a grandson, Patrick, who was born the year this book was released. Seeing my son and his partner parenting in the most generous fashion makes me so joyful! The babies burble as they chase each other up and down the stairs at home, every day teetering on the edge of disaster while being simultaneously wonderful. I love playing with them and making them laugh in very silly ways.

With my own kids, descending into the world of baby chaos for the first time was a huge step for me. Being obsessive, I bought a kilometre-long roll of brown paper. I covered the kitchen floor with the brown paper every day so the food they mysteriously put anywhere and everywhere except their mouth was easier to clean up. But the mess from the new babies just doesn't bother me now. I fully accept that the massive external disorder they create is more than compensated for by the even more massive internal order they manufacture daily inside their bodies and brains – and the joy they bring. I even deal with my special zebra-striped paper-clip collection, which used to be both size- and colour-coded, getting tossed all over the room and then returned in absolutely no order at all – I have mellowed, I guess!

I'm incredibly lucky that my kids and extended family are still happy to include me in their lives and share their relationships, stories, thoughts and experiences with me. They love me, despite all the fumbling and missteps that came with me learning to be a parent. It makes me so happy to be with them and to listen to and learn from them.

I wish I had been more generous to my parents with my time, given how happy it makes me to be with my own children. You live and learn, and it took a long time for me to grow up emotionally. I would like to think I'd do things differently with my parents now that I have more insight into the love they held for me. Having children opened my eyes to the multiple ways my parents did everything they could to make my life better. Sometimes the things that parents do for you are hard to bear, especially if they see part of their role as telling you what you are doing wrong! But my parents went out on a limb and took risks to try to help me, knowing that I might not be receptive. I now see their attempts were intended to help turn me into the best possible version of myself.

I can't do anything about being a better son at this stage of my life, given both my parents are dead. But I am going to keep trying to be a better daddy (or daddy daddy, as my little granddaughter Gemma calls me. I am a little surprised that she already understands mathematical recursions at such a tender age – not even two!).

My family – including my extended family – are the very core of my heart and my life, and I can't imagine being without them. They make everything in my world more joyful!

For as long as I can, I will offer my kids and family all my love and support. I'm very proud of them and I love them all to pieces.

1987–90 and 1990–2010

Outback odysseys

I have lived most of my life between Wollongong and Sydney. Like 86 per cent of the Australian population, I am based in one of the eight big lumps that we call capital cities.

About 40 per cent of Australia is so dry, with rainfall so low, that it is counted as desert. Another 40 per cent is dry enough that you can't really grow crops on it. That leaves only 20 per cent of the country as potentially agricultural. It was different in the past, before colonisation. The landscape has changed enormously since the rabbit, livestock and unsympathetic Western agriculture were introduced almost 250 years ago. In some areas, two metres of topsoil has gone after 60,000 years of cultivation, while other areas that used to be grasslands are now sandy deserts.

Despite my city slicker status, I became obsessed with wanting to see the Australian outback at the end of the 1980s. At the start of our outback adventures, we travelled for the love of it, but later it turned into a paid gig, which made it even easier to keep on doing what we loved!

I began by buying every four-wheel drive magazine I could find on the newsstand. I wanted to learn as much as possible about the sort of vehicle we would need to get us through the Australian outback.

Toyota LandCruisers were the usual choice, but not for me. I ended up going down a totally eccentric pathway with an ex-military Volvo C304 that I modified to suit our needs. The advantage of sticking with what everyone else drove was that you could walk into Fred's Friendly Fruitorium and pick up the exact part you wanted. Even though the Volvo was going to be harder to get fixed while we were away, seeing everything was geared towards LandCruisers, I wasn't worried. I had all the mechanical know-how and plenty of spare parts.

The Volvo had a top speed of 80 kilometres per hour, but it was virtually a 'go anywhere' truck. Its raised axles gave us incredible ground clearance, and it had loads of usable space in the back for all our gear, 300 litres of petrol stored in onboard tanks, 200 litres of water, a pop-top tent roof, and 300 watts of solar panels. I even fitted a hot shower on the side.

It took about six months to fully modify the truck, then we did two short (just a few weeks and a few thousand kilometres) shakedown cruises – to find out what did, and didn't, work. That got us ready for our biggest trip, leaving from Alice Springs with baby Alice just recently born. The truck weighed in at five and a quarter tonnes, and nearly 900 kilograms of this was fuel, oil and water. We were heading for one of the toughest four-wheel drive treks in the whole world: the Canning Stock Route. For this trip we carried an extra 300 litres of petrol in drums to get us through the whole distance, traversing various deserts.

The Canning Stock Route came into existence because in the early 1900s there were huge numbers of cattle in the Top End of Australia, but at the bottom of Western Australia in Kalgoorlie, a meal with meat cost an ounce of gold. The East Kimberley cattlemen wanted to get tick-free cows from up north to the abattoirs down south – but they didn't have effective ways to treat the cows for ticks. In 1906, surveyor Alfred Canning was tasked with mapping a stock route and finding wells in the desert so the cattle could be moved overland nearly 2000 kilometres from the Top End. Mustering the cattle through the heat of the desert would kill the ticks en route, but the other problem was that no known wells were mapped. What Canning did was both dreadful and typical of the time. He didn't ask the local Indigenous groups where the water was – instead, he captured people, chained them by the neck, fed them salted food and waited for them to head to the nearest well to drink water. He did this over and over, across nine different Indigenous language groups. In this awful way he mapped the fifty-four wells that would allow the cattle to travel through the desert. The track was only used to transfer cattle a few times.

So, around October 1990, we headed west from Alice Springs for 1000 kilometres, then turned right and went roughly north for another 1000 kilometres through the Gibson and then the Great Sandy Desert. It was already getting hot.

Dr Karl Kruszelnicki

Oh my goodness, it was beautiful. We were lucky enough to come through the desert after rain, and everything was in bloom. There were yellow, purple and silver flowers and honey grevillea dripping in nectar. I used to suck the sweet liquid off the flowers like it was honeysuckle. There was so much variation in the vegetation as we drove along. People who have never been off-road might think the desert was all the same, like a 2000-kilometre stretch of beach. But for me, the desert was the opposite, delightful and varied!

Our pace was an incredibly slow 80 kilometres a day, with a long lunch in the middle. In that month of travel on the Canning, we saw only one other group of travellers. They were treating the trip like an endurance race. They drove from sunup to sundown and did not look happy! They no more than waved at us. On the other hand, we trundled from well to well, taking three days to get from one to the next. The wells were simply holes a few metres across dug in the ground, with a wood or metal cover that you could drag off the well when you wanted to pull up water in a bucket. Sometimes the water in the wells was crystal clear. Sometimes it was muddy and had dead birds in it – whichever it was, we drank the water! We didn't have enough cooking gas to boil it, so we purified the water by adding both potassium permanganate and artificially sweetened cordial – as well as helping to improve the taste, the cordial was also supposed to kill some germs.

Going west, you're running parallel with the sand dunes, and that's relatively easy driving. But once we hit Well 34 and turned right, heading north-east on the Canning Stock Route, we were going up and over every single dune, and that was definitely not easy going! While the Volvo was incredibly capable, it was not the ideal vehicle for driving on sand. We came across a sand dune every two kilometres, so there were five hundred of the buggers!

Sand driving is okay in a light vehicle, but of course we weighed about five tonnes. A powerful engine helps too, but we had a gutless six-cylinder, three-litre petrol engine. Big balloon tyres with no tread are the ideal choice, whereas we had aggressive mud and rock tyres with deep tread.

But we did have diff locks on each axle, and we did have massive ground clearance. Plus we had a shortwave radio with a range of thousands of kilometres if we needed an emergency rescue.

Driving on sand is a lot more work for the engine than a hard surface like rock. You might think that you're driving on 'flat' sand, but each tyre is always pushing up a small hill of sand in front of it. If your engine is not powerful enough, it will slow down as you climb up a steep sand dune until you just come to a halt – you're fighting the slope of the dune, gravity plus the little hill of sand in front of each wheel. But if you reverse back down the sand dune, you've got gravity on your side. Providing you go down in exactly the same wheel tracks as you went up in, you won't get bogged in the sand. Driving over the same sand like this compresses it, and compressing the sand makes it easier to drive on.

This was the technique that we used to get over the five hundred sand dunes. We would charge the sand dune at full pelt until we slowed down and were forced to a stop. Next we'd reverse downhill along the same sand tracks. Then we'd have a second go. And maybe a third go. It was slow, but ultimately we conquered each sand hill, one by one.

I'd calculated that 600 litres of fuel would be enough to get us from Alice Springs, over 2000 kilometres of desert, and back onto the highway at Halls Creek in Western Australia. It was a little tight. We left the desert to arrive at Halls Creek with only 40 litres of fuel left! With good planning and good luck, we didn't need an emergency fuel drop – which was lucky, because I think the cost would have been about $10,000. (An emergency fuel drop consisted of a 405-litre drum of petrol wrapped in a mattress and tied to the wing of a light plane. When they saw you stranded down below, they would cut the ropes and let the drum fall where it would – hopefully protected by the mattress.)

We were in the desert in late October when temperatures were reaching into the high forties during the day. Between us we were drinking about 40 litres of water a day. At night the temperature would drop to a much more comfortable 37 degrees. We got used to it!

Many of our friends were genuinely bemused as to why we kept on returning to the outback. But there was nothing dull about it. The Simpson Desert, for instance, has eight different habitats, each quite different from the others. They include grasslands, deserts with different coloured sands, saltpans, spinifex and so much more.

The astonishingly deep red of the Strzelecki Desert is quite different from the red sands of the Gibson Desert and Little Sandy Desert. It's so beautiful that it makes you weep! The

A Periodic Tale

Little Sandy Desert is (surprisingly) almost the same size as the Great Sandy Desert (about 110,000 square kilometres) and holds enough sand to cover all of Eastern Europe to a depth of two metres.

The Painted Desert in South Australia looks as though someone had loaded up some aerial tankers with different pastel paints and sprayed them on the landscape. And Victoria has its deserts in the west – Murray-Sunset National Park, Big Desert and Little Desert.

Our usual nightly routine was to have an early dinner and be lying down comfortably on groundsheets with a warm cocoa (or perhaps the occasional drinkie with an umbrella) by sunset armed with a laser pointer. The first ninety minutes after sunset were spent finding satellites (we'd use the laser to point them out to each other) and the next ninety minutes was spent looking for meteors (we'd find about ten each hour). Then we would go to bed. And of course, on moonless and cloudless nights, the glorious Milky Way was an amazing bonus. Many urbanites in Australia have never seen a satellite or a meteor. But thanks to the flatness of the outback, we saw them every night.

There was the period in late 1991 when, every night, we saw a strange purple band on the horizon near sunrise and sunset. Earlier that year, Mount Pinatubo in the Philippines had erupted over several weeks, the second biggest volcano event of the twentieth century. The dust cloud (all 10 cubic kilometres of it!) blasted 40 kilometres up into the stratosphere and then quickly encircled the entire Earth. The eruption killed nearly a thousand people, engulfed and closed the US

Clark Air Base, and reduced the Philippine economy by 8 per cent. Some sixteen passenger jets accidentally flew through the various dust clouds and sustained over $100 million in damage. In addition to the dust, 17 million tonnes of sulphate aerosols were injected into the atmosphere, and the dust and the aerosols set off a 'volcanic winter' that reduced the amount of sunlight landing on the Earth's surface by 10 per cent and briefly cooled the atmosphere by an enormous 0.5 degrees Celsius. In Australian cities, nobody noticed the thick purple band on the horizon – but we got to see it, for three whole months, while we were outback.

After leaving the Canning, we headed west-ish to the coast for another highlight of the trip, Derby, where I really loved watching the 12-metre tides. We stayed a while in Broome, which is like paradise, and then came slowly all the way down the west coast to Perth.

From there, Mary flew home with the babies while I drove all the way back across the Nullarbor. I picked up a hitchhiker for company. It turned out that my passenger, who in the dark looked like a little old woman, was actually a fit, wiry woman who spent the time in my truck telling me of her violent criminal activities. On top of that she showed me that her long, skinny bag held a samurai sword! The reality that she could possibly sever my head while I drove along, combined with her 'colourful' stories about crime gangs, was very scary. At the next well-lit road stop, I pulled in and told her that was as far as I was going (meaning that was as far as I was going with her!). She got out with no complaints. She seemed tough enough to look after herself.

Back in Sydney, I was filling the Volvo at a petrol station when the driver fuelling up next to me started chatting. People were always stopping to talk cars with me because the Volvo was a really unusual vehicle! (I had a similar experience driving in an old 1956 Chevy on Route 66 in the United States.) It turned out that he was an editor at Australia's most popular four-wheel drive magazine, and he was so impressed by my Volvo that he invited me to become a contributor.

As a result of that chance meeting, I wrote columns for a few decades and ended up test-driving four-wheel drives all over the Australian outback. We've seen practically all the coastline of Australia, fifteen of our seventeen deserts, and spent a total of some two years sleeping in the outback under the beautiful big night sky.

1990–91

Working up a storm

At the same time that we were having kids and driving around Australia, I started an unlikely chapter in my life as a TV weatherperson. This job started by accident, like a lot of things in my life. And it finished the same way – abruptly, on this occasion because I wilfully insisted that meat had fat in it!

Out of the blue, an old boss called me up who was now doing a TV brekkie show, *Good Morning Australia*. He offered me a full-time job as a TV weatherperson and I agreed, also offering to add some science segments. They got more content (22.5 minutes of science stories each week, to be precise – three 1.5-minute stories over five days), and I got more satisfaction!

The TV hours were as crazy as medicine – from 3 am till the end of the day. And TV land is tough, which after a while I found out for myself.

I picked up the skill of being able to present to an invisible TV audience without looking nervous. (That skill would eventually morph into me being able to speak confidently to any audience, physically present, or not.) I also got really good at reading the autocue while speaking naturally – the trick was to make it seem like I was not reading. Back then, the autocue was always in caps and sans serif. I hate sans serif typefaces! (Serifs are the little end bumpies you see here on the letters of the alphabet – they let you tell the difference between a capital 'I', as in India, and a lowercase 'l', as in love.) I printed my own autocue scripts in a serif typeface using both upper and lower case. This made it a lot easier for me to get the sense of what I was reading quickly.

I did a bridging course with the Bureau of Meteorology (BOM) to learn about the weather. Before that, I didn't even know that in the southern half of Australia, the weather (and the winds) came from the west to east – some weatherman! This explains why airline flights can sometimes be five and a half hours from Sydney to Perth (pushing into a headwind) but the flight back as short as three and a half hours (getting a free ride off a tailwind).

Weatherpeople don't make the weather forecasts themselves. Nope, it's the good old 'rip and read': the BOM sent the weather forecast out, and the media folk ripped it from the fax machine and read it on-air.

It took me a while to realise this, but in my experience, TV has a higher percentage of pathological liars – much higher – than any other field of human endeavour that I have come across (though I've no experience with politicians and real estate agents, as comparison). One time I asked about copyright ownership of pictures that I wanted to use on air. The person

I checked with told me to use anything I wanted – because copyright law didn't apply to live TV! This was a total lie, but I believed them because they spoke with such confidence. It was a big surprise years later when I found out that copyright laws absolutely do apply to live TV!

Another time there had been a newspaper story suggesting that women who breastfed were at an increased risk of breast cancer. I knew this story was just plain wrong and that breastfeeding in fact *decreased* the risk of breast cancer, so I said I wouldn't put out this story because it was incorrect. One of the producers got all excited and said, 'That's terrific! We can get double the bang – first, we tell it like it's true, and tomorrow we can run the story again and say that it's wrong.'

I responded, 'But what about the women who only see the first segment and stop breastfeeding because they wrongly think there is an increased risk of breast cancer, and maybe get breast cancer in twenty years by not breastfeeding?'

The producer's reply was an unrepentant and heartless, 'Fuck them.' I stuck to my guns and didn't touch the story. So that one didn't go to air, thank goodness!

I shared an office with hosts Tim Webster and Kerri-Anne Kennerley. I liked that they were both 'what you see is what you get' type personalities, a bit like the surgeons I knew. Our office was decorated with three large newspaper posters (called 'screamers') that newsagents put outside their shops to advertise the breaking news of the day. Each of the screamers carried the same three words, stacked one above the other: TIM WEBSTER FIRED. Curiously, they were dated years apart. I asked Tim what the story behind them was, and he said,

'Karl, on three separate occasions, I finished my morning shift at a TV station and drove home. Each time, it was only when I drove past my local newsagency and read those screamers that I found out I was fired from my own show. Three times!' Maybe he felt like a phoenix rising from the ashes every day, with a new job, despite the posters proclaiming he was out on his ear.

I started on what I thought was a reasonable salary, but one day I realised that my salary paled into insignificance compared to what the full-time hosts were getting.

I'd just bought an old Peugeot 405 from my Glebe car mechanic mate because our old car had died. He was asking $240 for the car. Sure, it had a few problems, but nothing major (except for a bad habit of stalling whenever you turned right in front of a truck!). In fact, when you accounted for the full tank of petrol it came with and the nine months of registration, the price of the car came down to just $40 rather than $240. I felt like I had got an amazing deal!

I drove in on Monday morning and proudly told Kerri-Anne and Tim, 'I got a new car for two-forty!' They dashed to the window and said, 'Which one?' I pointed out my second-hand bomb, and in unison they said, 'Oh, you mean two hundred and forty dollars!' The fact that they thought I was going to show them a car that I paid $240,000 for suggested they were in a very different financial league to me!

On one occasion, I nearly killed myself doing a live science experiment with helium.

Now, most people know that if you breathe in helium from a party balloon, it will temporarily make your voice go into a squeaky high pitch. So I thought I would explain why your voice changes as an experiment on TV. Actually, this is not a simple thing to explain at all!

I ordered a cylinder of helium gas, and about a minute before my science segment, I started breathing the helium in and out in deep breaths, straight from a tube hooked into the cylinder, until I got my voice satisfactorily high-pitched. Then I kept on breathing in and out really fast with the helium. I incorrectly believed that I could saturate my tissues in helium and keep the squeaky voice effect going for longer. I was running on the television rule that anything was okay – as long as the piece to camera looked good! But I didn't think this through properly – and it ended up being dangerous.

The cameras threw from Tim and Kerri-Anne to me, and I started speaking in the high-pitched helium voice to try to explain why helium had this effect. I had another suck on the helium partway through the segment, showing off for the cameras. I came to the end of my ninety-second explanation, then as the camera swung off me, I noticed that my thinking was getting fuzzy – and simultaneously, I began to crumple slowly to the floor. I was feeling odd, but I didn't realise immediately that it was because I wasn't breathing.

I collapsed completely, and made a bit of noise as I hit the ground. The camera operator nearest to me turned and put a finger to his lips, as though to say, 'If you're gonna die, just die quietly!' I was losing my colour vision and everything was going into black and white. My eyesight was closing down into a tunnel and I was heading into unconsciousness. (The retina is one of three very oxygen-hungry organs in the body, along with the brain and the heart.)

I still couldn't work out what was going on, but suddenly I thought back to my respiratory physiology lectures and remembered that you do not breathe in just because you are low in oxygen – you breathe in because you are high in carbon dioxide. When I had been hyperventilating, I had blown off all my carbon dioxide, so my body had no reflex drive to breathe. I fell to the floor because I didn't have enough oxygen to run my brain – because I wasn't breathing. In fact, my last breath of air was about three minutes ago. Oh dear!

I realised I had to take a breath, to avoid dying, and so get some oxygen into my lungs again. But it was really difficult to take a breath without the breathing reflex! But I kept on trying. Then suddenly, with a loud, gasping rasp, I took in a breath of air, and almost instantly, my consciousness snapped back to normal. It was like a light switch had been flicked back on! My tunnel vision opened wide to normal vision. As I lay there on the floor taking in loud gasps of air, the same cameraperson turned around and put his finger to his lips again!

There are at least three lessons in this.

First, doing experiments is hard. I hadn't done any special training in science demonstrations and it seems like I definitely should have! I usually now avoid doing experiments and stick to telling stories.

Second, every year in Australia, a few kids who are perfectly good swimmers die in underwater breath-holding competitions in backyard pools, while the adults aren't watching. The kids often hyperventilate before jumping down to the bottom of the pool, trying to saturate their tissues with extra oxygen so they can hold on for longer. But given the blood is normally around 97 per cent saturated with oxygen, another 2 per cent of oxygen would make no difference. Hyperventilating can in fact block the breathing reflex by expelling all carbon dioxide, as above.

So breath-holding competitions in the backyard pool should always be a no-no. It is just too dangerous!

Third, don't annoy the cameraperson.

Makeup mayhem

I first truly appreciated the expertise and skill of makeup artists when I turned up at the TV station one morning, and found the place crawling with cops.

Kerri-Anne was on-air for two hours each morning, from seven to nine. So her habit was to get up in the dark every morning, have a shower, put on her dressing gown, and drive the short distance to the Channel Ten studio in Ultimo to sort out the rest of the day.

On one predawn trip she was pulled over by the cops for speeding slightly. The cops checked her licence, and the first thing they said was, 'That's not you!'

Being a generous person, she said, 'Come along and watch the transformation.'

So the cops followed her to the TV station and went with her to makeup. Over the next few hours, they watched dumbfounded as she transformed from somebody driving a fancy car in their dressing gown into the glamorous daytime television star Kerri-Anne Kennerley. The cops had called in some of their cop mates to come and watch too, which is why, when I turned up around 4.30 am, I found the place completely littered with police.

Obviously, Kerri-Anne had 'good bones', but her natural good looks were enhanced by the talent and proficiency of the makeup artists responsible for her daily television transformation. I deeply respect their skills.

Based on their say-so, I also know that my skin goes best with a KT3 foundation base (back then), and I'm super happy to get Spakfilla plastered on my face before I go on-air. It just evens out my skin tones, dahling!

I was having a lot of fun in my new job and learning all the time. But after a year and a half, my TV weatherman career came to an abrupt end.

I had read a quirky tongue-in-cheek article in the *New England Journal of Medicine* about how to make a 'healthy' hamburger. The recipe first boiled the meat to remove the 'unhealthy' saturated fats, and then fried the boiled meat in 'healthy' monounsaturated olive oil at a low temperature. The patty was still deliciously crispy from the frying, but without the saturated fat. I organised a hamburger shop to make the burger and put up a sign advertising its New England Journal of Medicine Burger! It was a light and fluffy segment, perfect for breakfast television.

Straight after the show, my immediate boss told me to go see the 'suits' on the top floor. Apparently, the Australian Meat and Livestock Corporation did some $3 million worth of advertising every year with Channel Ten, and I had just gone on TV and said that meat had fat in it. That healthy wad of advertising cash was supposedly in danger! The funny thing was that my story was about how to eat more meat more healthily, but the suits didn't see it that way at all.

The proposed fix was for me to do a science segment the next day saying that meat didn't have any fat in it. Which I could not do …

And that's how my career as a TV weatherperson came to a meaty end.

1994–96

Julius Sumner Miller fellowship, horse racing and Erdős numbers

Sometimes the stars aligned and I found myself in the right place at the right time. I've often fallen into new jobs by just being there. That good luck, combined with my habit of talking to absolutely everyone, seems to have opened doors for me. And one of those doors was into an academic position at the University of Sydney in the School of Physics.

The physics school was set up in the 1950s at the University of Sydney, with Professor Harry Messel as its first head. Harry had a big beard and a gravelly voice, was larger than life and was widely recognised for saving the crocodiles in northern Australia. He had also been a student of physicist Erwin Schrödinger, a pioneer of quantum mechanics and famous for his Schrödinger's Cat thought experiment. Harry was still around, but no longer head of school.

Dr Karl Kruszelnicki

Quantum fuzzy

Werner Heisenberg laid the foundation for quantum mechanics. Without quantum mechanics, your phone wouldn't work.

The famous Heisenberg uncertainty principle states that with subatomic particles, you can know either their exact location or their exact speed (momentum) – but not both at the same time. This is not because of our inadequate measuring instruments but is a fundamental property of the universe. Yes, down at the tiny scale, the universe is fuzzy. And it essentially means that an electron can be in two places at the same time!

This Heisenberg joke is so bad that I'll have to explain it for you. Heisenberg is driving a car and is pulled over for speeding. The cop says, 'You were doing exactly, and let me emphasise *exactly*, 54 kilometres per hour.' Heisenberg replies in horror, 'Oh my God – exactly? That means I'm lost!' Ha ha! The 'funny' bit is that in quantum land if anyone knows their exact speed, then they can't know their exact location. And if you don't know your exact location, then you must be lost. You can have speed or location but not both!

> Schrödinger's Cat is a hypothetical thought experiment that builds upon this fuzziness.
>
> Suppose that a cat is inside a sealed box. You can't see into the box. Beside the cat, there is also a vial of poison gas which is on a random timer to break open and release the deadly gas. You don't know when it will open. The Big Question is, 'Is the cat alive or dead?' The Answer is that the cat is both alive and dead at the same time *until* you break into the box. But when you open the box, suddenly the cat becomes either alive or dead.
>
> Welcome to the Weird Science of Quantum, where even the old rule of 'i before e except after c' no longer applies. After all neither the word 'science' or the word 'weird' follows this grammar rule.

After I left *Good Morning Australia*, I'd been giving fun science talks for high-school students at the University of Sydney's International Summer School. Out of the blue, in 1994, I got a phone call from the head of the School of Physics.

The school was setting up a new position, the Julius Sumner Miller Fellow. (Professor Julius Sumner Miller was an American physicist who was a visiting lecturer at the School of Physics from 1963 to 1986 and a consummate TV performer.

His TV show *Why Is It So?* featured experiments and was filmed at the University of Sydney.) The call wrapped up with him asking if I knew anybody who'd be interested in applying.

I'm pretty concrete, so I just said, 'Sorry, I can't think of anybody,' and thought that was the end of it.

The next day, my fax machine pushed out a scrawled message: *Why don't you apply?*

I was totally astonished. I had never thought of becoming an academic, but life is long and strange. So I did apply, and blow me down, I got the job as the inaugural Julius Sumner Miller Fellow, starting in 1995, and I've been in that role ever since!

Being a Fellow is a lovely academic role – and loosely defined, allowing me enough flexibility to continue my other public roles. Ultimately, the uni wants to attract students to study science. By keeping up all the things that I was already doing on radio and television, and in magazines and books, I was able to help enormously with that overall goal. One year the uni did a survey of the students who had just enrolled to study science. One in seven said that I was the reason that they chose to enrol at the University of Sydney.

To further spread the Good Word about science, I give regular fun general science talks at the uni, and in country towns, for the public. At the end of each talk, I usually do a Q&A so the audience can ask me any of their burning science questions, and a major surprise to me has been the drop in the age of those asking the questions. Over the last decade, the average age has dropped to under ten – and the questions are becoming so deep and complex that they are beyond belief! These kids are amazing. I often see three generations of kids, parents and grandparents sitting together in the audience, but I don't know which generation brought along the other two! I also give more targeted talks to high school students (in person) about future careers.

A really fun aspect of my outreach work is giving science talks at comedy clubs, music festivals such as Woodford and Splendour in the Grass, and international venues such as the Royal Institution and the New Scientist events.

I also give two free science Q&A sessions with primary and secondary schools, over the internet, every Wednesday arvo. At this rate, in only a century from now, I will have done one with every school in Australia. I also do them with overseas schools. These sessions give me a great insight into what the students are thinking. I love actively mentoring the next generation of scientists, and teaching stagecraft to academics, so they can simultaneously use a microphone and enchant their students. It's a fantastic position and was tailor-made for me!

I got some great advice from the head of the School of Physics soon after I started as the Julius Sumner Millner Fellow in 1995. He told me that if I was ever invited onto an official government committee in Canberra, I should accept. You never know who you might meet. Blow me down again, but an invitation to join an innovation committee came my way shortly after, so I accepted. ('Innovation' is a vastly overused word in politics!)

Flying to Canberra early one morning, and being a friendly and curious bloke, I introduced myself to the person next to me and asked what line of work he was in. When he replied that he was a mathematician, quick as a flash, I asked, 'What's your Erdős number? And what's your favourite equation?'

He paused, looked me up and down and said slowly and deliberately, 'Of course, the Euler identity has to be my favourite, and my Erdős number is one.'

I couldn't stop myself blurting out, 'Erdős number of one? Bullshit!' Luckily, he laughed!

Paul Erdős (Erdős pronounced 'air-dish') was one of the most prolific and original mathematicians of all time. He slept only three hours a night and did mathematics seven days a week, nineteen hours a day, until he died at the age of eighty-three. He believed that a mathematician was a device for turning coffee (and amphetamines!) into mathematical theorems.

Erdős loved only mathematics. He once said, 'I cannot stand sexual pleasure. It's peculiar.' He didn't care about property, food, clothes or paying taxes. He never learned how to prepare or cook food, although he could add milk to breakfast cereal. His only possessions were some old clothes and a couple of battered suitcases. He lived by the old Greek saying that the wise man has nothing he cannot carry in his hands. He was a gypsy mathematician, who never stayed anywhere for more than a month. Erdős was like a bee, flitting from flower to flower, to cross-pollinate maths.

His social and practical skills were virtually non-existent. On one occasion, his hosts had left a window open in the kitchen where he was working. Before sunrise, a very heavy rain began coming in through the open window. Erdős woke his friends and said, 'It's raining in through the window. You need to do something.'

When he visited you, the deal was simple: he would bless you with the fruits of his brain, and the price was doing an 'Uncle Paul sitting', which meant taking care of his physical needs like money, transport and doctors' appointments to get him more amphetamines!

Yes, he lived on amphetamines. At one stage, a fellow mathematician told Erdős that he didn't need amphetamines and could do brilliant maths without them, so he reluctantly gave up amphetamines for a month. But he found he couldn't do maths, so he started back on the amphetamines while admonishing his friend, 'Do you know what you've done? You've put the progress of mathematics back by a whole month!'

And here I was, sitting next to someone telling me they had an Erdős number of one. Erdős numbers are determined by the degree of separation you are from having worked with Erdős – it's like the mathematical version of 'Six Degrees of Kevin Bacon'. You got an Erdős number of one only by writing a paper with Erdős himself. That was a really big deal and a badge of honour! If another person then wrote a paper with you, that would give them an Erdős number of two. And so on.

Only about five hundred mathematicians have ever had an Erdős number of one. Gavin Brown, the man I found myself sitting next to, was in a very exclusive club. He must have been a mathematician at an incredibly high level. (When I checked this recently, Gavin Brown is indeed listed as having a low Erdős number – but of two, not one! I dunno.)

I ask all mathematicians I meet for their favourite equation and their Erdős number. Aside from Gavin Brown, two others really made an impression on me, even though neither had a low Erdős number! Instead, they both had great maths tattoos.

One had the famous equation known as Euler's identity tattooed on her forearm – so cool! Imagine loving an equation so much, that you are prepared to wear it forever. I don't have any tattoos, but if I did, it would probably be Euler's identity.

Euler's identity

Euler was a Swiss mathematician of the eighteenth century who, like Erdős, was another of the most brilliant mathematicians ever.

The famous Euler's identity is a lovely equation, as follows:

$$e^{i\pi} + 1 = 0$$

In this one simple equation, you have three mathematical fundamentals.

First, there are the basics of counting in the numbers 1 and 0.

Second, there are the basics of algebra in the addition symbol and the equals symbol.

And finally, there are three of the fundamental constants of the Entire

A Periodic Tale

> Universe! There's the irrational number e, which is about 2.717, and is the basis of logarithms and exponential growth. There's π (or pi), another irrational number, which is basic to understanding circles and spheres. Finally, there's i, which is the square root of minus 1 and the most fundamental of all imaginary numbers (because in reality, no number multiplies by itself to get a negative – but that's another story!).
>
> All of those beautiful numbers in just one tiny equation. I love how jam-packed but absolutely simple Euler's identity is!

The other mathematician who impressed me had a different forearm tattoo:

$$y = |fn(x)|$$

I looked at it, but I didn't get it – y is a function of x? She pointed out that the vertical lines each side of '$fn(x)$' meant 'always take the positive value of the answer'. I suddenly got it: her tattoo meant 'always be positive'. Wow – life advice from maths!

Anyway, Gavin and I had a lovely chat on the flight from Sydney to Canberra about how wonderful doing mathematics with Erdős was (OMG, Erdős!), and many other things. Then we disembarked and went our separate ways. An hour later

I walked into the meeting room in Canberra, and guess who was chair of the committee? Gavin! It turned out that he was also the vice-chancellor of the University of Adelaide (which he hadn't bothered telling me, because an Erdős number of one, or two, was much more glamorous!).

This chance meeting on the flight to Canberra led to Gavin saving my bacon about six months later.

My initial Julius Sumner Miller fellowship appointment was only for twelve months. In late 1995, the new head of the school called me in and said, 'We love you very much. But we've got no money to pay you. You can stay without pay or you can go – your choice.'

I went back to the office and panicked for a while. My academic career was about to come to a grinding halt after one short year. Then I suddenly remembered that Gavin Brown was now the vice-chancellor of the University of Sydney and that we both loved Erdős.

I made an appointment and went up to his office. I told him I was screwed and was about to lose my income! Did he have any advice? He said something like, 'Karl, as vice-chancellor, I've got access to a few hidden logs of money. I'll shift forty thousand across to cover you for a while.' Then he gave me a list of other people inside the university (outside of physics) to whom I could offer my services for funding. If I got work with them, as well as with the School of Physics, that could be my way out. He finished with, 'And by the way, never ask me for money again – this is a oncer.'

I was so grateful. His advice was spot on – the $40,000 got me over the hump, and various other schools in the university chipped in to pay for my now ongoing position as a Fellow.

Gavin fixed my problem in only three minutes – great time management! In the twelve minutes left in our quarter-hour appointment, he told me a story about how Erdős had accidentally got him into $15,000-worth of trouble!

Gavin had a hobby, aside from being Numero Uno at the University of Sydney and a mathematician. He spent every Wednesday afternoon jaunting off to the horse races – and would win. How? The winning was based on his high-level belief that all of the racing industry in New South Wales was to some extent corrupt. Based on his belief, he reckoned the horse races were fixed. In other words, it was already known *before* the race which horse would win. All Gavin had to do was work out which horse that was.

He didn't care about the names of the horses, or their past performance or parentage. He just looked at the numerical pattern of the odds fluctuating on the betting board. Straight maths! Based on his belief that racing was crooked, he reckoned that the 'smart money' would, at some stage, land on the horse that was going to win the race. But the people in the know would have to do it subtly; they couldn't swing the odds too much or too quickly.

Gavin was a professional mathematician, and this was a good test of his skill (and his Erdős number of one – or two!). He focused his big mathematical brain on the numbers dancing across the odds board, until he saw a pattern that he recognised. He would bet on that specific horse and would consistently win a lot of spare cash.

Paul Erdős himself was his honoured guest for a month, so he organised a dinner. Of course, all the invited mathematicians turned up. But during the dinner, in the middle of an animated theoretical mathematics discussion, Erdős wandered off to check a theorem. He returned shortly after with a maths book and a fat bundle of cash in hand. 'Gavin, I found this inside the book,' he said, and handed over $15,000. To Erdős, the money was just an inconvenience slowing him down from finding the theorem he wanted. To Gavin, it was his emergency stash. Gavin had hidden it in an obscure maths book, somewhere he thought nobody else would look, not even his wife. That got him into trouble at home!

It's funny to think that I started as a medical student on a little patch of land at the bottom end of Sydney Uni in 1981. I left for a while and then swung back full circle by taking the Julius Sumner Miller fellowship in the School of Physics just a few hundred metres away! Since then I have just stayed put, and my little team has had an office in the School of Physics for years now. I really fervently want to thank everyone who has worked with me over the years, for all their creative ideas and enormous support.

For me, the best part of my job is the wonderful people I meet in the corridors. Sure, the big-name academics are often eccentric and entertaining, but I really love the vibrancy of the clever students and their ways of seeing the world in a different light.

In the School of Physics, I also love listening to visiting academics and non-academics from elsewhere in the world.

One of these folk was in the physics department corridor looking a bit lost one day, and as it turned out she was looking for Dr Karl Kruszelnicki. 'Jackpot,' I said. 'You've got me.'

'I really like this part of my job,' she said in an American accent. 'I'm here to offer you distinguished international foreign visitor status with the United States of America.'

I said, 'Wow,' and offered her a cup of tea. Then I asked, 'What exactly is a distinguished international foreign visitor?'

She told me the US government was offering me an all-expenses-paid one-month trip to see anything I wanted to, anywhere in the United States, no strings attached.

What an amazing offer! Quick as a flash I said yes, followed by, 'Can I go to NORAD?'

'Definitely not!' she spluttered.

I stuck out my bottom lip and looked very sad and downcast (I have kids, after all, and this trick often worked for them to get their way with me). Eventually, she relented.

'Well, okay, maybe,' she muttered.

And that is how, somehow, I ended up being authorised to visit the North American Aerospace Defense Command (or NORAD) headquarters inside a hollowed-out mountain in Colorado.

The role of NORAD has been wound down quite a lot in the years since. But back then it acted as the ultimate surveillance machine, continuously gathering and coordinating military information. The dual purposes of this facility were to assess all incoming information about a potential attack on the USA but also to remain functioning in the worst-case scenario – say, even after nuclear weapons had landed.

The NORAD complex could survive a 30-megaton nuclear explosion detonated as close as 2 kilometres away!

The idea that I could go somewhere so secure and off limits, and with such high technology, really appealed to me, so the following year, off I went to Colorado. Driving close to NORAD, I passed signs that read 'Authorized Visitors And Staff Only' and 'Deadly Force Can Be Used' (a euphemism for 'we can legally kill you'!). I came around a corner to find the road blocked by a tank right in the middle of the road, and my first thought was, 'Oh my God – are they going to kill me right now?' Luckily I was waved past the tank, and I parked the car.

I was provided with my very own personal guide/guard who stayed with me all the way through the visit. He was much bigger than me and apparently some kind of martial arts champion in the marines. I think he was there in case it was necessary to overpower me if I suddenly went stark raving mad, or maybe in case a major event compromised the security of the USA and Canada and the Cheyenne Mountain complex had to go into lockdown. (I later found out this was called 'button-down mode'.) During any such conflict, of course, they'd shut the big NORAD front doors and I'd be safely locked inside with them, away from the rest of the world.

My other wish for that trip was to see and touch a Lockheed SR-71 'Blackbird', a plane so fast it could fly from Sydney to Perth in just over an hour.

At Edwards Air Force Base in the Mojave Desert, my dream came true and I sat on the front seat of an SR-71. I can still make airline pilots on commercial flights insanely jealous by showing them a photo to prove it's true! It was such a deeply moving experience for me, as impressive as being inside Notre Dame or York Minster cathedrals. The SR-71 was a superb example of technology and human labour making something that was absolutely the best of its time.

The only reason I got this trip of a lifetime was because I chatted with an American engineer who liked all the same science and technology as me. He had very high-level connections in the US government, though I didn't know that when I met him. I just did what I always do and started a conversation.

In my experience, you just never know where a conversation will lead you. It is so worth talking to people and finding common ground.

2003

The god of overcoming obstacles and new beginnings

Despite my hippie past, I had never been to India.

So when my son, Little Karl, wanted to go with his Year 10 classmates on a school excursion to India and Nepal over December 2003 and January 2004, I was very keen to go along. Luckily for me, the group of boys needed a few parents for 'crowd control'.

The trip included travel to Sikkim and Darjeeling in India, and to Nepal. One of the other parents (who was herself born in India) clinched my decision to join by relaying an old proverb: 'If you see the sun set in Sikkim, you can die happy.' In a worst-case scenario, it seemed a reasonable ambition. Little did I know that our trek would turn out to include the worst and best – both a near-death experience and a marriage proposal!

India has nearly 1.5 billion people, some twenty-two official languages and half a dozen officially recognised religions (and dozens of other religions) – and despite that diversity, it's somehow avoided all-out civil war (so far)!

After the birth of Christ, India was the largest economy on Earth for about 1700 years, but then the British used India's wealth to fund their many wars. They plundered India so completely that the Indian economy dropped from about 35 to 40 per cent of the world's economy to just a few per cent.

Sikkim is a state in the far north-east of India and borders Tibet, Nepal, Bhutan and West Bengal. It's very small (only 65 by 116 kilometres) but extremely mountainous – so much so that it would take you about three months to trek the 65 kilometres across Sikkim! If you were to trek the same distance but on flat ground, it would take around three days. Previously an independent state, Sikkim 'voted' to become part of India in 1975 (still a contentious issue for some Sikkim inhabitants). It's also had cross-border 'interactions' with China in 1962 and 1967, and today, armed Indian and Chinese infantry are still just 40 metres apart in sections of the border.

Suze, the school's art teacher, was leading the tour. She was a very fit yoga practitioner and surfer who had been to India many times, and was energetic, positive and always supportive. At the start of the trip I knew her only very casually, but that was about to change dramatically!

We landed at Kolkata airport in the middle of the night, and I vividly remember the absolute shock of the new. As a government school group, we were definitely travelling on a tight budget, trekking and carrying all our possessions in backpacks. Our first mistake was to pack all our penknives, torches and handy tools in the outside pouches of our backpacks. All these useful gadgets had vanished by the time we picked the backpacks up from the luggage carousel!

The terminal was so run-down that we had to pick our way around the broken ground in the parking lot. A memory burned forever in my mind is of a small child of maybe five years begging, while holding a tiny baby in her arms – at around two in the morning! She didn't talk to us, but just kept stretching out her hand and putting it to her mouth. Both of these little children should have been in bed, like my own five-year-old daughter back in Sydney, not in a parking lot in the middle of the night, begging for money from strangers. The contrast between my daughter's life and theirs just brought me to tears – and we hadn't even left the airport's parking lot!

It was a heartbreaking introduction to a country that is so rich and wonderful in many ways and absolutely shocking in others.

The smells of India hit me immediately. There were always five strong undertones in differing ratios depending on where you were: flowers, sewage, food, car pollution and smoke. Everywhere we went, people got rid of their rubbish by burning it, usually at the end of the day. Each year, about 70 million people on planet Earth die, and about 10 per cent of them die early from breathing polluted air. Burning anything creates polluted air!

Another eye-opener was that everyone in India could spit very skilfully. While Europeans generate a big blobby spit, Indians eject a long, cylindrical, well-aimed, slow-speed projectile. But skilful or not, spitting is a powerful way to spread infectious diseases.

Most of us got pickpocketed. Sometimes, even the zips on our inside pockets were undone! Soon we all carried our valuables in zipped pouches under our clothing. The local thinking seemed to be that it's okay to pickpocket, but it's not okay to use weapons to steal somebody's possessions.

In India, I would see men holding hands with men, and women holding hands with women, but never men and women holding hands. Hand-holding implied friendship, pretty much the cultural opposite of adults in Australia at that time, when it was mainly women and men who held hands publicly to imply a romantic link.

Then there were the crowds! At Kolkata station we were immediately surrounded by beggars, ranging in age from three-year-olds to adults. Kids were grabbing on to us and not letting go! It was all very new and confusing.

From Kolkata we headed north for Darjeeling in the foothills of the Himalayas on the overnight train, another strikingly new experience. Even the classes of train travel were completely unknown to me, but they ranged between: AC1 (air-conditioned first-class, with four-bed compartments and curtains); AC2 (air-conditioned second-class, six beds with thin cushions and no curtains); sleeper (no air-con, six beds, thin cushions, no curtains); seats (non-tilting, closely packed); and floor. I had never seen tickets for floor seating on trains before! We were in relative comfort in first-class.

Near the border of Sikkim, we changed to a bus. The road was narrow and steep, with terrifying 200-metre sheer drops and only a low, crumbling concrete wall between the road and the abyss. All the trucks had signs on the back saying 'SOUND HORN', meaning to beep when overtaking. And this was one rule that everyone *did* obey!

Our bus driver had wild habits, including very unconventional and dangerous overtaking! When he wanted to overtake a slow-moving truck, he changed down to a low gear to get more power. So far so good. But then, just as he drew alongside the truck he was trying to pass, he would shift up to a high gear and instantly lose engine revs, power and speed. Instead of sailing past the other truck and moving back safely to our side of the road, we would just trundle along side by side – but with us on the *wrong* side of the road – as we went around blind corners, above breath-holding cliff edges! I understood instantly why the drivers kept an infinite number of statues and pictures of multiple gods stuck on the dashboard of the vehicle. The only thing saving you from a head-on collision or a plunge over the cliff edge had to be divine intervention.

Looking out the bus window, I would often see a half-built building, with concrete floors laid down and rusty reinforcing steel rods poking out of the concrete at crazy angles. There might be little thrown-together shanties on the concrete, with people living in them. This turned out to be a sign of astute financial planning, and not of decay. If a family got extra cash, they would immediately do more building to the family home rather than see the money frittered away. So the houses were built bit by bit, every time the money trickled in.

We climbed into the foothills of the Himalayas and came to Darjeeling, home to the prince of teas. While only 10,000 tonnes of Darjeeling tea are grown each year, some 40,000 tonnes of something *called* Darjeeling tea are sold. I bought some genuine tea, and it genuinely tasted amazing! Since then, I have kept up the daily Darjeeling habit – but now I order it online to make sure it's the real deal.

From Tiger Hill on the outskirts of Darjeeling, we had the thrill of seeing Mount Everest with our naked eyes! It took our collective breath away. It wasn't too cold where we were, but Everest was snow-topped. It looked like a regular mountain, but taller!

After a few days in Darjeeling we drove the 100 kilometres over four to five hours to Gangtok, the capital of Sikkim.

That very sunset we also saw Kangchenjunga, the third-highest mountain in the world, in all its magnificence – even though it was 53 kilometres away. In fact, a few weeks later on our trek, we got to see four of the five tallest mountains on Earth all at the same time! They were Everest (8847 metres), Kangchenjunga (8586 metres), Lhotse (8516 metres) and Makalu (8485 metres). The only one in the top five that we missed was the K2, the second tallest mountain.

After Gangtok, the entire group headed into the serious trekking. It took a few weeks, with nightly stops in tiny villages where we slept in the houses of locals. Early on we had porters carrying our backpacks who listened to the cricket all day on little portable radios. One pair of porters had the perfect system: one would carry the body of the tiny transistor radio while the other kept the tip of the metre-long antenna in his mouth, so his entire body acted like an antenna! This effectively made the antenna two or more metres long, giving significantly better reception.

As part of the trip, the school kids volunteered on projects, like painting stupas for Buddhist monks. Stupas are roundish shrines in various sizes that are often used for meditation. Little Karl (who by now was already 1.87 metres tall and matched me in height!) was a group leader and organised us into teams. A typical job was to prepare the surfaces of the little stupas for painting, or to chip the old concrete away to allow a new layer to be placed on top.

One day we carried over a tonne of water up the hill to the concreters. Little Karl loved maths, and it seemed to me that he had mentally organised a time-and-motion study, mapping out the jobs in the most mathematically efficient way. As soon as one job finished, we flowed automatically into the next job, with no wasted time.

We walked from temple to temple, which was an amazing experience. A lot of the Buddhist monks were only kids themselves, and up for mucking around. They loved playing cricket with an almost religious fervour, and I wondered if their parents were sending the kids to the monastery to learn to behave properly.

One temple sign that I loved told us that it was a 'Notified Religious Place of Worship. Killing, Drinking, Smoking & Intoxicating Within This Compound Is Strictly Forbidden'. So it seemed smoking and killing were on a par!

Some temples had dozens of red-and-gold prayer wheels lined up for visitors to turn. Prayer wheels are spinning drums for Buddhist devotion. We saw a golden temple bell 3 metres high. The main building in one temple complex was three storeys high and absolutely amazing in terms of the religious art, so rich and diverse!

There is a wide variety of plants in the Himalayas, some of which are poisonous. Traditionally, elders passed on important knowledge about plant poisons. We were warned very seriously not to touch or pick any flowers. In one area we trekked through, the locals insisted we wear bandannas as face masks because a plant released chemicals that had killed people, cows, dzos (a cross between a yak and a cow) and dogs! (From memory, I think it was a type of rhododendron.) It was like an excerpt from a Phantom comic seeing this straggly group of young boys wandering around with scarves tied across their faces. I'm not totally convinced that a bandanna would actually protect us from toxic fumes, but we all did as we were advised!

The terrain was also high enough for some of us to experience minor altitude sickness.

The science of altitude sickness

While trekking, I noticed that my breathing at 4200 metres was not really under my control but went into a weird cyclical pattern. Let me explain:

1 I would breathe deeply and quickly (maybe twenty times each minute) to compensate for the pressure being so low – almost 50 per cent less than the pressure at ground

level. This would let me get more air into
   my lungs and more oxygen into my blood.
2. But because I was breathing so deeply and
   quickly, I blew a lot of carbon dioxide
   out of my bloodstream.
3. Quite logically, I would suddenly stop
   breathing. (You'll remember from my
   daredevil-ish TV demonstration that the
   main 'driver' to take the next breath
   is not having a low oxygen level in our
   blood, but rather a high carbon dioxide
   level). Very low carbon dioxide meant
   I had a very low drive to take the
   next breath – so I didn't. Literally,
   I stopped breathing for about ten
   seconds. Unfortunately, my metabolism
   kept on running and using up oxygen, so
   the level of oxygen in my blood would
   also plummet.
4. Suddenly, with carbon dioxide building up
   again and very low blood oxygen, I needed
   to take a really, really deep breath,
   followed immediately by deep and quick
   breathing.
5. Repeat steps 1 to 4 (really annoying).

As we travelled deeper into the countryside, Suze started having back pain. She was stoic and didn't complain much, and her innate fitness was likely masking the symptoms to a degree. Her worsening pain was dismissed by the non-medically trained adults, who told her that once you hit the age of thirty, everything was inevitably downhill. But as the only medically trained person in our group, I knew that something bad was happening – I just didn't know what! It had been a decade since I last worked as a medical doctor at the kids' hospital, so I was really out of touch with adult diseases.

Looking back at photos, you can see Suze's posture becoming more bent over as the pain worsened. She told the schoolboys that it was just period pain (maybe scarring them for life if they thought that was what normal period pain was like!). She tried local massage treatments, but they just led to such excruciating intensification of her pain that she gave in to screaming in a way that terrified the local practitioners!

Then super-fit Suze's pain went over the tipping point until it was so severe, she couldn't even move. The pain meant that she couldn't walk at all anymore, so we made a plan to activate our medical insurance and get her picked up and flown to the closest city for treatment.

We soon learned that no helicopter would fly into a remote Indian mountain town bordering China, not when India and China (each with nuclear weapons) were bickering over 'minor shooting actions' between their soldiers at that exact time! Who knew that insurance companies had an exclusion clause for situations potentially involving nuclear weapons?

Her symptoms meant Suze and I had to leave the tour group.

I went with Suze and a driver by road to the nearest hospital to get both a diagnosis and treatment. We headed off in a Land Rover through the hilliest terrain imaginable, with snow and ice on the narrow dirt roads. It was beautiful, but I was too worried to look at the view much!

The Land Rover had 'issues'. First, literally none of the tyres had tread. Second, none of the four wheels were held on with all five bolts – apparently four was more than enough, and three was quite okay. Then the handbrake didn't work. Each time we parked (and we were always parked on a hill, because we were in one of the most mountainous regions on planet Earth), I had to put rocks in front of the tyres before the driver could lift their foot off the brake pedal. The back was open so there was nothing to keep out the sub-zero air outside, and the shock absorbers were broken, so the ride was very uncomfortable and painful for Suze and her back.

The dirt track was mostly only one and a half cars wide and carved into a very steep hillside, with snow varying between just peppering the ground and up to a metre deep. When we met a vehicle going the other way, one of us would have to reverse until the road widened enough to pass each other, typically a hundred metres or so.

By this stage, Suze couldn't stand or sit. She lay in the open back of the Land Rover tray, and if we were going uphill I held on to her chin to stop her sliding out of the end of the tray onto the road. If we were going downhill, I kept a towel between her head and the back of the cabin to stop her bumping against it. At least when we were going downhill I wasn't worried about her falling out onto the road!

She kept her eyes shut all the time except once, when she opened them after a particularly violent swerving motion to see the back of the Land Rover actually hanging over the edge of the cliff! She wisely shut her eyes again.

We stopped at the first tiny town that had a pharmacy where, over the counter and without a script, I could buy opiates for subcutaneous injection. These medicines were seven times more powerful than morphine and cost just thirty cents each. I also bought a stack of sterile syringes and needles, even though the pharmacist said I could just wash one with water and reuse it!

I gave injections to Suze to reduce the level of her now excruciating pain so she could bear the bouncing travel. The good thing about the opiates was that they relieved her pain. The bad thing was that if I gave her too much, her breathing could slow down so much that she could die. And of course, the thin air at altitude contained much less oxygen than the air at sea level, so the margin of safety was reduced.

Each night of our exhausting drive, we shared a room in the tiny homes of kind local people who moved kids out of beds to free up space for us. Suze would have the bed while I slept on the floor. I needed to help her get up to go to the bathroom or do anything at all as her mobility was very badly restricted. I gave her painkiller injections whenever she poked me with the special stick she kept at her side to wake me! I didn't have any real way to measure the effect of the medicine I was giving her, nor her blood oxygenation levels, so I was trying to walk the fine line between keeping her pain under some kind of control and making sure she didn't overdose on the opiates. I was scared and exhausted!

Finally, after several days, in the late afternoon we arrived at a town with a hospital – but the CT scanner wasn't working because there was no electricity. We nearly cried. It turned out that there was no electricity because the town had no garbage collection. People would 'clean up' their rubbish by setting fire to it, and this created air pollution so thick that you needed electric lights to see. And to access electricity to turn the lights on, people were illegally tapping into the nearest power line with metal clips! Because so much electricity was being hijacked, there wasn't enough left to run the CT scanner.

The hospital had no running water, soap or toilet paper. The toilets emptied into a vacant lot next door, and the smell was pungent. There were nurses, but they had sold the right to look after the patients to local beggars, who then moved into the room with the patient. So I paid a local to look after Suze and left her overnight in the hospital while I found an el cheapo hotel to get a proper night's sleep for the first time in days.

Luckily, the phone in my hotel was reliable enough to ring Mary. Remember, I was now several days with hardly any sleep and stressed out of my mind, and at high altitude, so my brain wasn't running very well – but I needed some basic medical advice.

I got through to Mary as she was preparing breakfast and said, all in a rush, 'Hi, honey! You know how I'm in India with our son? Well, now he's somewhere else and I don't know where, but I'm with the red-headed teacher and I'm injecting opiates into her buttocks.' As if that wasn't enough,

A Periodic Tale

I went on, 'I seem to remember that I shouldn't do that. Am I supposed to inject somewhere else? Why don't we inject into the buttocks anymore, and where should I inject instead? Oh, and also, will you marry me?'

And then the line went dead!

By the time I managed to get another connection, I had forgotten all about the impromptu proposal (and I reckon that Mary didn't see this as a genuine proposal, and thought it was just part of my general hysteria). Over the course of the next few hours, I rang Mary back many times to try to explain Suze's situation. Mary was asking me a million sensible questions like 'Why?' and 'What are you doing?' and 'Do you know anything about what is going on?', as well as making reasonable statements like 'You're not even working as a doctor anymore.' All I could say was that there was no one else and I was it, before the phone cut out again! When I got through again, Mary suggested that injecting into the upper outer thigh was safest, to avoid accidentally hitting the sciatic nerve in the buttock!

Meanwhile, back in Sydney, Mary then had the stressful job of contacting Suze's family (whom she had never met) and explaining to them that their daughter was sick with some as yet unknown illness, that I was travelling with her and doing everything I could, but that I wasn't really a practising doctor anymore and definitely out of my depth. It was a fraught window for all.

Come the morning, the sun came up and electricity consumption plummeted, meaning there was finally enough to run the CT scanner. Suze and I were already waiting, first in line, as soon as it opened. And finally, thanks to the scan, Suze got the long-awaited diagnosis.

The scan showed a huge epidural abscess low down in Suze's spine. An abscess is a localised infection that needs to be removed by surgery, and usually won't respond to antibiotics. Suze's abscess was around the covering of the spinal cord in her lower back. On one hand, this perfectly explained her back pain. But on the other hand, this was completely unexpected, as Suze had absolutely no risk factors to make that likely – it was just bad luck. But at least now we had a diagnosis and were no longer in a disputed border area, so the medical insurance would kick in.

Our revised plan was to get back to Sydney for surgery, but first we had to get to Kolkata. There was no way we could keep driving, so we had to fly – immediately. Suze had a diagnosis, but she needed urgent treatment. The pain was getting worse, and I didn't want to keep cranking more opiates into her.

Even though everybody *wants* to help, organising anything in India is hard. Phones cut out, faxes (yep, it was that long ago) didn't send – and chaos reigned above all. There were literally dozens of phone calls to the insurance company in Sydney, the Australian High Commission in New Delhi, and the Department of Foreign Affairs in Canberra. The phone line repeatedly went dead at random times.

We needed a plane to fly from Kolkata to the nearby military airport, pick us up, and then fly us back to Kolkata. After a whole day, I had only managed to organise permission for the plane to leave Kolkata – but not to land at the military airport where we would be waiting.

Regardless, the next afternoon I loaded Suze into a van and headed for the airport. Amazingly, permission had arrived for the plane to land, and the military knew we were coming. (All those phone calls had paid off!) I had to go through a scanner. They let me keep my penknife but confiscated my matches!

The single-engine eight-seater rescue plane landed just before sunset and came with an emergency doctor, which was a great relief. We took off again immediately, and as soon as we hit cruising altitude, one of the first things the doctor did was measure Suze's blood oxygen level. It was a fair bit lower than expected – very low, in fact – and he gave me a stern look, which suggested that I had been giving more painkillers than he thought wise. Remember, opiates can reduce breathing rates and blood oxygen, and now we were flying at altitude in a plane, which can additionally reduce blood oxygenation to dangerously low levels.

With a big show to point out how precarious the situation was, he turned to the pilots and instructed them to drop to just 1000 feet, an extremely low flying height, especially in the dark. But that was the way we flew all the way to Kolkata.

The Kolkata hospital was actually amazing, very modern and efficient. It was nothing like the little rural hospital we had been struggling with just a day or so ago. In my mind, however, I was still thinking we should fly back to Sydney so Suze could get an operation to remove the epidural abscess at home.

In Australia, epidural abscesses are very uncommon. An Australian neurosurgeon might see one epidural abscess in their whole surgical career. In fact, while we were in India, a woman in Australia died from an epidural abscess that had not been diagnosed.

But this very hospital in Kolkata that we had landed in did many epidural abscess operations every week. In fact, it did more epidural abscesses than any other hospital in the world! The surgeon told me this was due to rampant tuberculosis infections in India, which somehow opened the door for bacteria to get into the central nervous system. What are the chances of arriving at the one hospital that considered epidural abscesses to be as common as bread and butter?

It was after midnight. The neurosurgeon had just examined Suze (neurosurgeons operate on any part of the nervous system, not just the brain), and he was brutally pragmatic. Sure, we could fly back to Australia for the operation. But that extra delay before Suze got surgery would likely mean she would permanently lose power and sensation in her legs, bladder and rectum because of the pressure from the abscess on the related nerves.

The alternative, which he was recommending, was to do the operation right there and then, at 2 am. In that case, the pressure would be removed from the nerves immediately, and the outcome would be far more favourable. Our call, he said – but there really wasn't much of a choice to make. Opting for immediate surgery over likely permanent neurological damage seemed like the only rational decision!

So in the darkest hours before dawn, we went ahead with the surgery in Kolkata, and Suze's outcome was really great.

By the time Suze was recovering well in hospital, her family had arrived to look after her. I was incredibly relieved that she had pulled through so well – and also very grateful that her family would take over her care. Her mother is a trained nurse and wonderfully practical, and her dad is such a kind and friendly guy that he treated me like his long-lost son. They wouldn't hear any of my apologies about how I wished I could have got her treated sooner. They were just so glad she was alive and well, and that I'd done everything I could, and stayed with her.

Suze did have pain for some time after the surgery, which she managed in her usual way with more exercise and fitness training. The only permanent damage was some altered sensation on the outside of her lower right leg, but she was alive and able to walk and continue with her active lifestyle.

We all became good friends in the aftermath of that baptism of fire. In fact, her parents put us up at their place when we're up their way – and anybody else in our super-extended family is also welcome to stay with them!

A few months after we all got back to Australia, there was a knock on the door, and there was Suze. She had commissioned for me the most beautiful surfboard I had ever seen, with a painting of Ganesha, the elephant-headed god, on it. Ganesha is now my favourite god – the god of new beginnings, wisdom, luck, and especially of overcoming obstacles. It was her way of acknowledging all that we had been through together.

But getting back to Kolkata, I was at the end of many nights without proper sleep. The ordeal was over. I was feeling almost delirious, but extremely conscious of the fragility of life. I had developed the strong desire for more meaning in the universe. I think the whole experience highlighted for me the need for continuity, ritual and symbolism. That was almost certainly why I had decided the best plan, straight off the back of this terribly stressful experience, was to ring Mary back home in Sydney and ask her a second time to marry me. For me, what persisted in my brain from that first frantic phone call was the sense that life was unpredictable, and that we needed to make visible symbols to mark what was precious!

She was so totally unprepared for that phone call and had thought that I would probably not even remember that I asked her to marry me a few days earlier, and that it was all part of my irrational crisis response! We had already been together for about twenty years by then, and had three happy and healthy children.

It was a huge about-face for me to want to get married now, because when I was growing up, I didn't think anyone should marry. As a teenager, the Cold War was in full swing. There were some 50,000 nuclear weapons on the planet, and just 1000 of those were enough to kill most of humanity. Back then, I didn't want to bring children into such a fragile and unsafe world, or set up a permanent relationship marked by a marriage. (Climate change brings a lot of this sense of impermanency and risk to some people today!)

The old hippie me also used to call marriage the 'M' word, and in many ways marriage had stood for a lot of the traditional religious ceremonies that I had no emotional connection to. I know Mary had never had any plans to get married either.

After getting back to Australia, I kept asking and my persistence paid off. In the end she said she would marry me, because she thought I was suddenly so keen to get married that I would have gone and asked someone else if she didn't agree. Of course, I wouldn't have done that, but I could see why she was concerned.

Suddenly, thanks to my close shave with a near-death experience, 'marriage' was our brand-new plan. As a bonus, it turned out that it made her mother very, very happy to finally be mother of the bride.

Little Karl and the rest of the school group had managed to continue the trip as planned, and so we all met up again in Kolkata. The school group visited Suze in hospital, and I showed them the before and after CT scans of her epidural abscess. None of the schoolkids had realised how sick she really had been, or how close she had come to dying! It was a little weird to be reunited, because the rest of the group had bonded tightly while Suze and I were off having our own stressful adventure.

When we all checked in at Kolkata airport for the trip home, I mentioned that I was with the school group. The check-in agent seemed to mentally change gears and very politely asked me if I was their teacher. I wasn't, of course, but to keep it simple, I said, 'Yes, I'm a teacher.' She then said, 'Well, I'll see if I can upgrade you to business class.' And she did! If only we Australians had as much respect for teachers ...

Something odd happened when we finally got through customs at Sydney airport to meet Mary and her mum, who had come to pick us up. I had grown a full white beard, as personal habits like shaving had gone by the wayside in all the drama of the medical emergency. Mary's mum saw me coming out of the gate and pointed me out, but Mary truly didn't even recognise me at all. Perhaps stress and worry dramatically downgraded my appearance – or perhaps it was the stupid white beard. In any case, I decided then to always dress up when in public, and never grow a beard again!

I've been back to India several times since, but never again have I had the transformative and intense experiences of that first trip!

2006

Getting hitched

Once Mary got her head around my proposal and eventually accepted my formal Offer of Marriage, we had to decide where and when to do it.

And, you guessed it, my idea wasn't just about rose petals and chocolates – I had a 'scientific' angle for our wedding.

It was almost a little embarrassing to be getting married at that stage of our relationship – it was not like we were spring chickens anymore. And obviously, a quarter of a century after meeting, falling in love, living together and having three children, it wasn't as though we desperately needed white goods or any of the usual blessings on our union!

So we opted to elope and save ourselves from explanations about why we had decided to get married *now*, after all the years we had been together. The upside of eloping was that we got to spend the cost of a regular wedding entirely on a trip to Norway to see the midnight sun, which was something we had both always dreamed of.

Why the midnight sun? I had read about it as a child, and it just seemed so unbelievable and exciting. After all, every day we can see with our eyes that the sun rises in the morning and sets in the evening, so how could it still be 'up' at midnight?

And why Norway? Well, the far north of Norway is above the Arctic Circle (which was essential to seeing the northern midnight sun), and Norway itself seemed both friendly and exotic.

So Wedding in Norway with the Midnight Sun was the number-one plan. My desire for the wedding was to combine romance, science and love. My dream of us getting married under the midnight sun was a symbol: in the same way that the sun would not set on our chosen wedding day, so I also hoped that the sun would never, ever set on our love for each other.

It turned out to be easy to choose, but hard to organise.

Getting married in a foreign country involves a decent amount of paperwork, which we didn't realise at the start. (Yep, fools rush in where angels fear to tread.) We spent a long time running between government offices in Sydney getting various documents certified in order to go ahead with the plan. But after a few months, the length of time it was taking to get the last few Norwegian official documents was plain ridiculous. So we eventually let our fingers do the walking and phoned the relevant government departments in Norway after finally narrowing our problem down to just one recalcitrant bit of paperwork that we couldn't make sense of. There were two forms, translated into English, that looked identical to us, so we couldn't work out which one to complete.

We spent a lot of time on long-distance hold trying to sort it all out, and the Norwegian government's official hold music was 'Fly Me to the Moon' by Frank Sinatra. We were always ringing in the middle of our night and giggled hysterically with both tiredness and anticipation (of getting married, or at the very least of finishing the paperwork) every time we heard that song. Going to the moon was a long-term dream of mine, but the midnight sun in Norway was closer and the perfect compromise.

When we finally got a human on the line, I explained our problem with the seemingly identical forms to him.

Completely deadpan, he said, 'Aha! You, Karl, are you a man or a woman?' I replied that I was a man. The follow-up question was, 'And you, Mary, are you a man or a woman?' When Mary replied a woman, he said, 'So it seems you need the Heterosexual Wedding Form B, not the Homosexual Wedding Form A.' Hah – so simple!

You see, back in the 2000s, foreigners coming to Norway specifically to get married were usually gay, because it was one of the very few countries in the whole world where this was legally allowed. In fact, way back in 1993, Norway was the second country in the world, after Denmark, to provide official government recognition of same-sex couples via a registered partnership. This was not officially a marriage, but provided virtually all the benefits, responsibilities and protections as a 'traditional' marriage. In 2004, a legal bill was proposed to abolish the registered partnership and simultaneously make marriage laws gender-neutral, so that people of the same sex could officially marry. That bill was passed into law in mid-2008, making Norway the sixth country in the world to legalise same-sex marriage.

But Mary and I were applying to get married in that intermediate window before the gender-neutral marriage law had been enshrined in legislation and while same-sex couples were still coming to Norway to get hitched under the registered partnerships legislation. Being a foreign heterosexual couple made us the odd ones out, which was a nice reversal of stereotypes, I guess! But it meant we did not fit with the standard foreign gay wedding that the Norwegian officials could organise in a heartbeat.

Once we sorted that out, everything else fell into place – we just had to book the flights. It is actually a really long way from Sydney to inside the Arctic Circle. Our four back-to-back flights were Sydney–Singapore, Singapore–London, London–Oslo (outside the Arctic Circle) and Oslo–Kirkenes (inside the Arctic Circle). Travelling continuously, it took us about forty hours. We looked pretty rough by the end of the last leg!

We were getting married on the summer solstice, the day with the most daylight hours of the year, which at the Arctic Circle meant the midnight sun would be in the sky for the whole twenty-four hours. My lovely daughter Lola had been totally convinced that I was lying to her about the sun never setting above the Arctic Circle during the time of the summer solstice. 'Daddy, you're not making sense,' she defiantly said. 'The sun has to set at the end of the day. It can't stay up in the sky!' She was always very firm about what she thought was right and wrong in the world, and in her rules, the sunset happened every day. The midnight sun simply didn't fit into her knowledge base.

> **What causes the seasons?**
>
> Both the seasons of the year and the midnight sun happen because the Earth is tilted at an angle of about 23.5 degrees relative to the sun. Basically, at the equator there is always twelve hours of daylight and twelve hours of dark. At Sydney where I live, about 34 degrees from the equator, the amount of daylight varies during the year from about nine and a half to fourteen and a half hours a day. But once you get beyond 66.5 degrees from the equator, which is the Arctic Circle in the north and the Antarctic Circle in the south, you can get twenty-four hours of sunlight in summer.
>
> Even university graduates can get this simple astronomy and physics wrong. In 1988, Matthew Schneps of the Science Media Group at the Harvard-Smithsonian Center for Astrophysics made a short documentary film asking Harvard students, graduates and staff one straightforward question: what causes the seasons? Out of the twenty-three

people asked, twenty-one gave a wrong answer, with most thinking it was because of a variation in the Earth's distance from the Sun. Only two got the correct answer, which is that seasons are related to the tilt of the Earth and the Earth's orbit around the sun.

Only two! And at Harvard! So don't blame the general public for not knowing, and especially don't blame eight-year-olds.

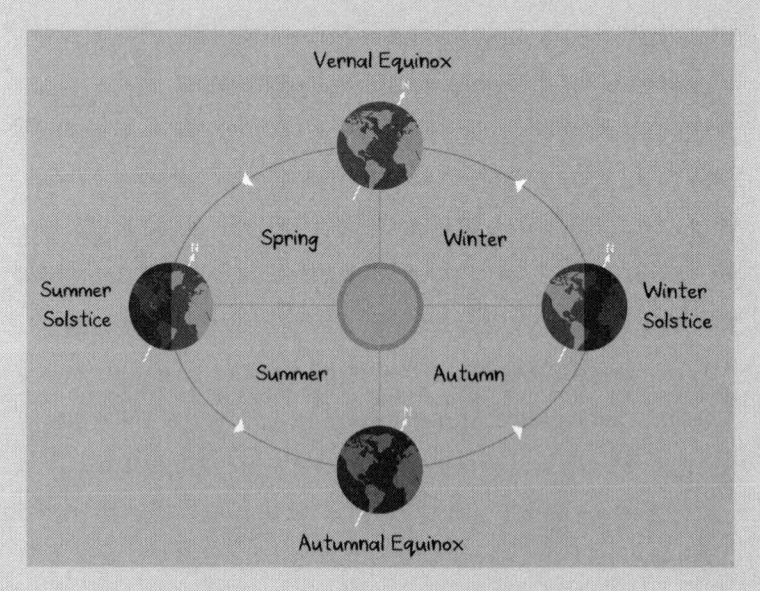

The northern summer takes place when the northern hemisphere is tilted towards the sun.

The wedding itself had to be in a church because that was the Norwegian rule. We chose an old church on the border of Russia and Norway for the Big Day. It was very picturesque, with misty tendrils of fog trailing around the striking copper-topped stone building.

Being us, we didn't elope by ourselves. We brought along our two daughters, Alice and Lola, and Mary's parents, Max and Carmel. Our son was invited but was off watching some football match instead! (Okay, it was the World Cup in Germany with his slightly older uncle Brendan, and that combination was a big deal! For him, there didn't seem to be any difference to his life if we were married or not married.)

My pre-wedding meal was a breakfast of champions – herrings, beetroot and coffee. It was the happiest day of my life, and not just because of the fabulous breakfast!

We intentionally tried to avoid the standard trappings of a wedding. First off, after brekkie we asked a random person in the foyer of the hotel to take the photos. (Hey, so the tops of our heads are missing from some of the photos, but the price was right!) Second, our wedding car was a purple and very functional four-wheel drive truck. Lola, then eight, was quite clearly unimpressed, but it was a really practical vehicle in all the mud. Third, we stopped our purple truck on the way out of town and bought a bunch of flowers from the *blomsterbutikk*. We chose a simple but pretty bouquet.

The church was partly divided between those of us who cried and those who didn't. I, of course, was on the crying side with my older daughter, Alice. We have always shared an emotional ability to laugh and cry at the very same things, even though in many ways I send Alice insane with my structured approach to life in general! Mary and Lola were on the non-crying side, along with Mary's parents, a church organist and a couple of tourists who seemed entranced by our random wedding. That made up the totality of the congregation. My parents were of course both dead by this stage, and being an only child meant there were no other blood relatives for me to invite.

The minister's wedding speech was supremely silly. I am pretty confident the priest had used the primitive 2006 version of Google Translate to write the sermon. We almost got the giggles halfway through!

I imagine what he was trying to express was that because we were getting married after all these years together, our love must have bonded us very strongly, and the marriage was a sign of great optimism, and hope that our love would continue into the future. There were hints of the film *Captain Corelli's Mandolin* about the sermon, with images of the roots of our trees joining together and getting stronger with time, but the translation of his words into English did not quite reflect this subtlety. Indeed, it was outrageously brutal and hysterically memorable in the end.

I swear he actually started by saying that maybe I used to look good as a young man, but now I looked more like a sack of badly packed potatoes. Not much hair, either. And Mary? Well, she might have been pretty when she was younger, but the blossoms had already dropped from her face and the roses in her cheeks had now turned to chalk! And that was before he started moving on to talk about love and desire. In summary, because we were by now so old as to be almost decrepit and physically incompetent, we must truly love each other deeply, because it could no longer just be lust!

That wedding speech gave us something to talk about for a long time to come. We invited the pastor to our wedding lunch anyway, in a private room of a restaurant, which was lovely. Some wonderful friend had sent champagne – but without a card! So we never worked out who to thank for the gift, but we did enjoy every drop. Reindeer and carrots featured on the menu, from memory.

And it was with great joy that we stayed up all night. With her own eyes, Lola could see the sun move down close to the horizon but never sink below it as it skimmed around us in a complete 360-degree circle. Lola finally understood that the midnight sun was real, because she had witnessed it with her own eyes.

So there we were, Officially Married! The next step was our honeymoon – with my in-laws and our beautiful daughters.

First, we jumped onto a ship and bounced our way 1500 kilometres down the Norwegian coast, a fantastic trip. It was almost as though there was no actual solid coast, but just islands, bays and fjords continually interlocking with each other. The picture-perfect scenery made you feel like the whole of Norway was an advertisement for health and nature! It was just so incredible.

The rest of the honeymoon was guided by a rough attempt to trace some of my parents' past experiences. After all, we were just around the corner from where I was born, though Helsingborg turned out to be totally underwhelming. With no links to any people, we just wandered around, wondering if my parents had ever walked on these streets with me as a little baby.

Then we went to where my mother had told me she was born: Gdansk, in what was now Poland, though it has been claimed by many different countries over the years. It was a nice seaside town with a great tradition of strolling after the evening meal. As you know, it turned out my mother was not born in Gdansk at all. I found this out thirteen years later when I participated in the television documentary *Who Do You Think You Are?* The research done by the producers for that show taught me more about her life than she ever told me herself.

Travelling to where my father was born was better, because we had organised a bit of a guided tour and he had told me more accurate details about his past. Over the last thousand bloody years in Europe, Poland morphed from a minor power into a great power before being totally swallowed up by its neighbours, and it even stopped existing as a single country for 123 years. It was reborn as the Second Polish Republic in 1918.

My great-great-aunt Salomea Kruszelnicka (also Solomiya Krushelnytska, depending on which alphabet you use, Cyrillic or Roman) is our family star. According to Wikipedia, she was a 'Ukrainian lyric-dramatic soprano, considered to be one of the brightest opera stars of the first half of the 20th century. During her life, Krushelnytska was recognised as the most outstanding singer in the world.' In our family lore (and backed up by Wikipedia), she moved in with Puccini to help him rewrite his opera *Madama Butterfly* to suit her voice, making it the grand success it eventually became. She was also supposedly the only woman to reject Puccini's amorous advances!

Salomea was born in a tiny village called Biliavyntsi in Ukraine, and is buried in Lviv. My father told me that as children, on a holiday at the beach, she buried him up to his neck in the sand! In her old apartment in Lviv there is a whole museum dedicated to her career, and I got a strong sense of my family connections from visiting it. The displays were full of photos and family trees that gave me links to that lost side of my family. After the museum we went to her grave, which was quite splendid and imposing.

Tired but happy, we returned to our life in Australia as husband and wife. My son was right: getting married made no difference to our daily routines.

Mary was still making my trademark bright shirts. My love of bright clothes goes back to my childhood. I have a memory from back then of a woman in a bright dress, standing like a lighthouse, in a sea of grey and brown on a miserable, rainy day in downtown Wollongong. She left a wake of smiles behind her as she walked by, literally and figuratively brightening people's day. So bright clothes for me seemed a shortcut on the journey to joy.

Before Mary was sewing 'one-offs' for me, I wore a lot of Hawaiian shirts, but synthetic fabrics get smelly really fast. To get a cotton business shirt with long sleeves and two pockets in bright fabrics was a task for my own personal dressmaker! Mary buys whatever fabric she likes, and I get to wear them once they've been sewn up as my 'stage clothes'. For me, the hope is that my colourful shirts might brighten other people's days too, just like that lady I saw as a teenager in Wollongong!

One big bonus of being married was that we now had easy labels for each other. We didn't have to stumble with introductions explaining how we were related, when meeting new acquaintances!

While the whole point of the wedding was to celebrate and create ritual around our love for each other, simplifying the terms of endearment was an unexpected perk.

Epilogue

It took me a long while to 'come of age' – way past leaving school or turning twenty-one. Frankly, there's still a lot for me to work out!

I've tried multiple career paths and been to uni many times over; a whole sixteen years of my life have been spent happily studying at various universities. I fully accept that people do things at their own speed and that it's okay to take time to 'find your feet', and your niche, and your people.

One thing I still struggle with is getting more in touch with my feelings and reading the feelings of others. I am so very impressed by today's young people, who seem so much more connected with their emotions than I was at their age – or at any age in my life! That emotional intelligence is something that I know can have a great impact, and I value and long for it.

The most intuitive emotional step for me was getting together with Mary and finally marrying her. It made me feel complete. Our marriage was a consolidation of everything up until that point for both of us in our personal, family and creative lives. Loving our own three children – and their partners, and their children, and our immense extended family – has given me a connection I never imagined possible. The tiny walled-off nuclear family of my mum, dad and me was so different from where I have landed today as a grown-up, totally immersed in family.

The two decades since our wedding have been wonderful. Having little babies in the family again (as the next generation) is unexpectedly exciting. It's so fun to watch babies transition from not being able to crawl, to being able to crawl only backwards and getting themselves stuck under a chair waiting to be rescued, to finally working out how to crawl forwards and then even how to walk. I get so surprised at how well they can communicate from such an early stage. Way before they are able to speak clearly, they can still definitely tell you what they want! And then once they get words, the interactions just blossom. Babies are not a zero-sum game, because babies seem to generate happiness and love out of nothing – repeatedly!

In terms of my ongoing work, it's been a mixture of 'same same but different'.

It's interesting how the various media themselves are changing. For example, one big 'but different' thing is the fact I somehow have more than half a million followers on TikTok. The audiences for newsprint and television are shrinking, maybe permanently, and even the definition of science on television is changing. After several years on one commercial TV station, my weekly science spots were 'replaced' by two people – an astrologer and a palm reader. I was flattered they needed two people to fill my shoes, but totally confused about why the network went straight to pseudoscience.

That's the 'different,' but the 'same same' includes my regular ten science shows on ABC radio as well as other frequent appearances on commercial radio and TV. With my Triple J gig, I think I hold the honourable title of Oldest Person on Youth Radio in the Known Universe. I have also continued to write books (this is number forty-eight) and journal articles. Just this morning I completed rehearsals for the highlight of my acting career – *Play School* on ABC TV.

I'm still at the University of Sydney, and still trying to learn more every day from the clever people around me. My role as a Champion of Science is fluid and allows me to be creative, curious and flexible.

'Same same' also includes still trying and failing at new careers! I ran unsuccessfully for the federal Senate in 2007 on a climate change platform. It started because one fateful day, as I was yelling fruitlessly at the TV, I suddenly decided, without any discussion with the family, to put my hat in the political ring. I didn't realise that meant I would have to resign from all my regular paid gigs at the university and the ABC. Luckily for me, both organisations took me back when I sheepishly returned with a record number of votes 'under the line' but no seat in the federal Senate. It was a huge learning curve, and a great (but expensive) education. I am now absolutely zealous about encouraging people with integrity to enter politics – 'P' is the first letter of both 'politics' and 'power', but that's not the only thing they have in common.

As far as bringing about change, I have no real power as an individual, but I do have some influence. I'm happy to use whatever influence I can bring to the table to inform the public about issues like climate change, which is already impacting the lives of people around the world! And at least in my role as a science journalist, I can present facts without toeing any party line.

The 'but different' also includes half a dozen trips to Antarctica. On one of them, I did the first ever science talkback radio show from Antarctica!

As a family group we have walked some 800 kilometres across Spain on El Camino de Santiago, one of the world's most famous pilgrimages. Among other things, I learned that not everyone who claims to be on the path to enlightenment is telling the truth.

Just for fun, I built six Foucault pendulums! That was 'same same', in that it's the kind of thing I would create, 'but different' because I'd never made one before. A Foucault pendulum is the only way to prove that our planet rotates. You set up the experiment down here on the ground and can see the proof within a few hours of watching the pendulum's movements (sorry, flat-earthers!). The best and final model was hung for several months in the central atrium of the Queen Victoria Building in Sydney. It was really beautiful to see it swinging back and forth, with people watching mesmerised on every level of the building.

Further on the planetary front, I have my own (minor) planet! That's certainly 'different'. Called Asteroid 18412 Kruszelnicki, it's about six kilometres in diameter and somewhere between Mars and Jupiter. A fan who also happens to be a very famous Scottish-Australian astronomer, Robert McNaught, named it for me as an honour. So even though I have so far failed to get to space myself, my little asteroid is out there travelling as my proxy.

I'm still giving science talks, except now I also do them at music festivals. I have even moonlighted as a go-go dancer on stage (including once at the Apollo in London, as part of a Brian Cox Science Variety Show!). The enthusiasm of the crowds is invigorating to me as a performer.

I also did a science entertainment tour for the Australian troops in Afghanistan.

In 2019 I was awarded the UNESCO Kalinga Prize for the Popularisation of Science.

The way I promote my books certainly is 'different'. I briefly held the Guinness World Record for the number of books signed in a single day by an author. The traditional cheese-and-wine book launch isn't my cup of tea; in 2007 I did a *literal* book launch. Yep, I launched my book out of a rocket on Bondi Beach on live TV! We got an ex-military rocketeer (who now runs their own space rocket company in Australia) to do the mechanics. The publisher's lawyers were especially worried about the low-flying book-rocket entering the restricted air space for commercial planes, but the Laws of Physics are stronger than the Laws of the Land, and the physics/mathematics said it was safe. One of the lawyers actually said, 'But what if the laws of physics are wrong?' The book launch successfully blasted off with a whoosh, and no aeroplanes, seagulls or people were harmed in the process!

While I didn't know it at the time, the early years of my meandering careers were never 'fillers'. Practically everything I learned was essential to where I ended up.

'Being there' was important – in the sense of 'being alive'. When I started my first professional job at the steelworks, many of my new colleagues had been there for twenty years. Some had lived one year of life and repeated it nineteen more times, but some had lived twenty different years of life. I learned the lesson of 'right here, right now' – living each moment and not just going through the motions.

I always wanted my work life to be overwhelmingly enjoyable. I have succeeded so much in this goal that I choose to 'work' all the time, because it's so much fun that I never want to stop. I do not stick with things mindlessly, and whenever my work life has become messy, I would change things – which often meant a new line of work (and occasionally being fired)!

But I also appreciate the argument for the exact *opposite* of 'loving your work'. In one of my various careers, I became friends with a storeman, whose job was really not difficult at all. Very quickly, I realised that he was blindingly intelligent – much smarter than me. He had deliberately chosen a job that would never stress him or make him anxious, with clearly defined hours and that also paid well enough. His joy was his life *outside* the job. And luckily, I have a lot of that too.

I could have written many different versions of my life for these chapters, and in the multiverse of my life there are plenty of other storylines. But hopefully I've picked some of the more interesting themes for this book!

For example, if my parents had enough money for tennis lessons, I could have been a better player (not just the under 16s, division 2, Milo Cup champion of Wollongong). Or maybe, I also could have been a champion high jumper. In 1964, at the age of sixteen, I realised that the then popular technique of high-jumping (scissors jump) was wrong – from the point of view of physics. Yes, you had to get your centre of gravity over the bar, but you didn't have to also have your trunk and head above the bar at the same time: your head and trunk could sail over the bar at a different time. The Fosbury flop (introduced by Dick Fosbury at the 1968 Olympics) could have been the Kruszelnicki kick, and I could have had an Olympic gold medal. But I didn't follow through by doing the hard training.

If I had followed my early curiosity about ceramics, I might have (maybe) been able to get a Nobel Prize in physics for superconductivity, rather than my somewhat less impressive Ig Nobel Prize in 'Interdisciplinary' Research (yep, it was misspelt on the award!) when in 2002 I was briefly a World Expert in Belly Button Fluff – where it came from, and why it was almost always blue – which Harvard University properly recognised. Painstakingly my colleague Caroline Pegram and I collected samples and, knowing the old rule that 'Anything, no matter how boring, always looks better under an electron microscope', we made history. All of my travel and accommodation costs to collect my prize (a set of wind-up chattering teeth on a stick) in Boston were *not* covered by Harvard University, because they did not want to insult my personal integrity with mere money.

And there are other what-ifs, too. What if I had accepted the guidance and help that was offered to me by the teachers in Film and Television School? Could I possibly have directed an Oscar-winning film?

And what if I (Olympic gold medallist and Nobel Prize and Oscar winner!) could have convinced Media Magnates to take a different approach to climate change in their newspapers?

I'm very open to immortality and living as long as I can — with a healthy mind and healthy body. I want my life to be driven by curiosity, generosity and optimism. But all things come to an end. Currently this includes a mortal life and thankfully also autobiographies!

So in keeping with the Frank Zappa song with the screaming electric violin solo that I woke up to many days of my life (try sleeping through that opening!), this book comes 'Directly from My Heart to You'.

Acknowledgements

The thought of writing my own autobiography has filled me with dread – which has not eased at all even now the book is finished. I still feel a deep sense of anxiety about what parts of my life story would be of use, or interest, to anyone else. And I really hope that I have chosen well, for the reader's pleasure!

Which all means that this book has been a long time in the making and would probably never have arrived if the Lovely Book Publishers at HarperCollins had not set me a final deadline. The deadline was the one thing that I recognised as familiar when comparing writing my usual science books to my autobiography, and for me nothing is ever done unless it is done at the last minute. So, thank you to all my editing and publishing team for your energy and encouragement and enthusiasm and perseverance in bringing this project to completion.

I have deliberately decided NOT to mention most people by name in this memoir, except friends who have died. This was partly driven by foolishly having not kept a diary, and so having to rely on my unreliable memory for details from the past. (Anonymity allows a little wriggle room if my memory is inaccurate.) And there are so many people in my life who are, or have been, influential, and it would be completely impossible to name them all!

Some people I do have to name individually because they have always been absolutely there for me – even when 'being there' included being brutally honest with me.

First, Mary (my wife), who is most essentially the love of my life and the person around whom my heart revolves. She may joke that my work is my first love – but that is not true. She always has my heart. And on top of that she is also the very best editor I have ever had. Writing needs editing and I run all my written words past Mary, because she loves the power of the big red pen and knows how to make the work really flow. As my Ghost Writer, she knows me very well and she can often say what I mean better than I can put it myself!

Second, my clever and adored children (Karl, Lola and Alice, in order of birth) have also edited and made additions to the storyline in this book – some of the better punchlines are likely from them!

Third, the Australian Broadcasting Corporation (ABC) has been my media 'work-home' over the last forty-five years. Kind and patient skilful people have helped train me in various technical and creative skills – in radio, TV and, by default, writing. They include producers Chris Norris, Tony Barrell, Daniel Driscoll and Nick Gerber, and all the Very

Clever Technical people in Master Control Radio, Sydney. You know how much I love you all.

And then of course, all the Triple J presenters who have hosted me in our fabulous long-running Science Q&A radio shows, as they get younger and I get older! Over the decades they have included the inaugural show with Angela Catterns (1990–1996), then Jen Oldershaw (1996–1997), Sarah Macdonald (1998–1999), Francis Leach (2000–2002), Steve Cannane (2003), Mel Bampton (2004–2007), Zan Rowe (2008–2017), Linda Marigliano (2018–2019) and Lucy Smith (2020–).

Fourth, the University of Sydney has been surprisingly welcoming and supporting of an odd bod like me. I started regular ongoing work as a Fellow in the School of Physics in 1995 and have found infinite wisdom from the staff working there. I love the way that physicists have the goal of understanding the world around them. I learn so much from them all.

Fifth, I've always worked in a small team, and my 'back room'– first Caroline Pegram and now Isabelle Benton – have kept me up with my crazy workload. They are both very creative, efficient and good at finding the optimal path. Caroline was the driving force behind making my Ig Nobel Prize in Belly Button Fluff happen, and Isabelle did the same for the UNESCO Kalinga Prize for the Popularisation of Science. Zoe Vaughan at Claxton's has been incredibly wonderful as my speaking agent for practically ever.

And lastly thanks to every single person who has ever genuinely wanted to know the answer to a question – and thought that I might be able to help to explain it to them in a straightforward way!